MYTHEMATICS

MYTHEMATICS

SOLVING THE TWELVE LABORS OF HERCULES

Michael Huber

Princeton University Press

Princeton and Oxford

LIBRARY OF CONGRESS CATALOGING-IN-PUBLICATION DATA
Huber, Michael R., 1960-
Mythematics : solving the twelve labors of Hercules / Michael Huber.
p. cm.
Includes bibliographical references and index.
ISBN 978-0-691-13575-5 (alk. paper)
1. Problem solving. 2. Heracles (Greek mythology) I. Title.
QA63.H83 2009
510–dc22   2009008535

British Library Cataloging-in-Publication Data is available

This book has been composed in Adobe CaslonPro

Printed on acid-free paper. ∞

press.princeton.edu

Typeset by S R Nova Pvt Ltd, Bangalore, India

Printed in the United States of America

1 3 5 7 9 10 8 6 4 2

I DEDICATE THIS BOOK TO MY HEROES—

to my father, Erwin Huber,

and in memory of my mother, Maria Huber.

CONTENTS

LIST OF FIGURES

FOREWORD

Introduction

Hercules (or, Herakles, as he was known to Greeks) is a hero well known for his strength and ingenuity. According to *The Oxford Classical Dictionary*, he shared the characteristics of, on the one hand, a hero (both cultic and epic) and, on the other hand, a god. He is arguably the greatest and most famous of the classical Greek heroes. The mythology surrounding Hercules has been a part of human culture for over 2500 years. Many students get a glimpse of Hercules long before they go to high school or college. Whether it is from a civilizations history course, a literature course, or watching a Disney movie, most of us know who Hercules is and we sometimes wonder if such a hero really existed. A walk through a famous museum, such as the Louvre in Paris or the Metropolitan Museum of Art in New York City will reveal an abundance of vases, sculptures, and paintings of Hercules performing one of his famous labors. Did you ever wonder how he accomplished them?

I decided to research the Twelve Labors of Hercules to develop problems in a wide variety of mathematical settings. What jumped out at me was the notion that many of the more famous labors of Hercules could fit into my mathematics courses. I have therefore endeavored to develop problems with varying levels of mathematical skill. I also hope this becomes a source book for both students and teachers to link the interdisciplinary aspects of mythology and mathematics. Some of the tasks that I created for the labors seem to have a nice fit with differential and integral calculus, both single-variable and multi-variable. Others require some understanding of ordinary differential equations, introductory matrix algebra, statistics, or physics. Still others just need a bit of logic or geometry. Mathematicians build a toolbox with the various mathematical tools at their disposal. My purpose became one of applying the use of these tools, motivating mathematics through applied word problems. I adapted applications problems in calculus to fit some of the Hercules scenarios. Not all of the labors give rise to great mathematics problems, however. Some are trivial, and in those instances, I used a bit of poetic license to create what I hope are more interesting problems. Others are not so trivial, such as when Hercules wrestles with Antaeus (in the Eleventh Labor), and they require a better foundation in material beyond calculus, such as differential equations and Laplace transforms. In digging through the myriad of Hercules information available, I was able to discover quite a few problems from about 2000 years ago buried in the writings of ancient Greek writers, such as *The Greek Anthology*, a collection of epigrams, poems, and riddles. Several of the mathematical puzzles were written down by Metrodorus, who was probably a grammarian living in the time of Constantine the Great (306 to 337). There are 150 such problems and puzzles, and supposedly, they "can be easily solved by algebra."

To convey the story of Hercules in an accurate manner, I consulted as many sources as possible, using library reference books, online web pages, personal conversations with classics scholars, and even Hollywood sources (yes, I watched many old Hercules films, plus Ray Harryhausen's *Jason and the Argonauts* (1963) and *Clash of the Titans* (1981), hoping that the fantastic special effects might provide some insight—okay, I just like watching those two movies...). The *Perseus Project* (see the Bibliography) is a great place to start if you want more information about Hercules or other classical Greek mythological legends. However, as one classics professor told me, "the myth of the labors of Herakles is just that—a myth," which accounts for the struggle to obtain any consistent facts surrounding the stories. Hence the name of this book: *Mythematics*.

There are at least four authors from classical times who wrote in significant detail about Hercules and his labors: Apollodorus, Diodorus, Hyginus, and Euripides. Each of these four authors tells the tale of Hercules and 12 labors, but not all are the same. For more details on the authors of the Hercules myths, see Appendix C. I chose the Apollodorus version, as his treatise, *The Library*, is deemed by many to be the authority. The work quoted in the following chapters is taken verbatim from Apollodorus' *The Library*, from the English translation by Sir James George Frazer. Apollodorus, in Frazer's translation, uses "Hercules" instead of "Herakles", so I will, too.

To set up the story, we need to understand why Hercules performed the labors. The following is taken from Apollodorus:

> Now it came to pass that after the battle with the Minyans Hercules was driven mad through the jealousy of Hera and flung his own children, whom he had by Megara, and two children of Iphicles into the fire; wherefore he condemned himself to exile, and was purified by Thespius, and repairing to Delphi he inquired of the god where he should dwell. The Pythian priestess then first called him Hercules, for hitherto he was called Alcides. And she told him to dwell in Tiryns, serving Eurystheus for twelve years and to perform the ten labours imposed on him, and so, she said, when the tasks were accomplished, he would be immortal.

In the legends, Eurystheus assigns various labors to Hercules, and in each, he does not expect our hero to succeed. Each of the labors is created as an apparently impossible task. In some, the jealousy of the goddess Hera reemerges to put Hercules to the test. Only a man of superhuman or godlike abilities can accomplish the labors. Hercules is that man. Hopefully, in transforming these problems into mathematical models, you, the reader, are also up to the task.

Format

Mythematics is laid out in a simple fashion. I list each task at the beginning of each chapter, in case you wish to read the task and solve it on your own before reading my solution. Each of the Twelve

Labors begins with a description of the labor quoted verbatim from Apollodorus. Then I provide a short synopsis about the tasks needed to complete each labor. This synopsis is then followed by the specific problems to be solved. In the second part of each chapter, I offer each task with a smaller pertinent quotation from Apollodorus, and then I give a detailed explanation and solution method for each one. Most labors require two or three tasks, involving different levels of mathematics. For example, the first task, dealing with the Nemean lion, has tasks involving velocity components, tilings (tessellations), and probability. You can choose to solve one task or all of the tasks for a given labor. In some of the chapters, I have included extra problems called Exercises. Hercules was commanded to perform the labors, but in doing so, he often had to accomplish additional deeds or exercises. The problems associated with the exercises are above and beyond those needed to perform the labors and hopefully will give you extra challenges. A complete listing of the mathematical subject areas for each labor can be found in Appendix A.

Acknowledgments

I would like to thank several people for their help. First, I must thank Vickie Kearn at Princeton University Press, who listened to my ideas and was totally supportive from our first meeting. Equal thanks go to Anna Pierrehumbert, my associate editor. Both Vickie and Anna have been extremely positive, motivated, and helpful throughout this project. Thank you to Dimitri Karetnikov and Anne Karetnikov for producing the figures in this book. Wow! I especially appreciate the comments and suggestions from John Quintanilla and Tim Chartier. I am grateful to Laura Taalman from Brainfreeze Puzzles, Inc., for the sudoku puzzles at the end of the Eighth and Twelfth Labors. Special thanks go to Jen Fellman and Lisa Perfetti for their assistance in translating the editor's comments from a 1776 Leonhard Euler paper which I used in the Ninth Labor. *Merci!* I also would like to express my gratitude to my colleagues at Muhlenberg College, especially Bill Dunham, Penny Dunham, and Dave Nelson.

I have dedicated this book to my parents; they truly have always been my heroes. I also want to thank my wife and family. To Terry; to

Nick, Tabby, and Riley; to Kirstin; and to Steffi: your encouragement
has been inspirational. Let's all continue to live happily ever after.

In closing, I would like to pass on two more quotations. The first is
attributed to the aforementioned Sir James George Frazer. Sir James,
in addition to translating Apollodorus' *The Library*, was a noted classics
scholar. He wrote,

> The longer I occupy myself with questions of ancient mythology,
> the more diffident I become of success in dealing with them, and
> I am apt to think that we who spend our years in searching for
> solutions to these insoluble problems are like Sisyphus perpetually
> rolling his stone uphill only to see it revolve again into the valley.

The second quotation comes from mathematician Paul Halmos and
is the opening line from his book, *Problems for Mathematicians Young
and Old*:

> I wrote this book for fun, and I hope you will read it the
> same way.

I believe that the sentiments of Apollodorus, Sir James, and Dr. Halmos
apply to studying and teaching mathematics. Hopefully the problems I
have created here are not insoluble and you will have fun solving them,
enjoying better success than Sisyphus; however, I offer no guarantee
that when the tasks are completed, immortality will be yours. Have fun!

MICHAEL HUBER
Allentown, Pennsylvania

MYTHEMATICS

The First Labor: The Nemean Lion

From Apollodorus:

First, Eurystheus ordered him to bring the skin of the Nemean lion; now that was an invulnerable beast begotten by Typhon. On his way to attack the lion he came to Cleonae and lodged at the house of a day-laborer, Molorchus; and when his host would have offered a victim in sacrifice, Hercules told him to wait for thirty days, and then, if he had returned safe from the hunt, to sacrifice to Saviour Zeus, but if he were dead, to sacrifice to him as to a hero. And having come to Nemea and tracked the lion, he first shot an arrow at him, but when he perceived that the beast was invulnerable, he heaved up his club and made after him. And when the lion took refuge in a cave with two mouths, Hercules built up the one entrance and came in upon the beast through the other, and putting his arm round its neck held it tight till he had choked it; so laying it on his shoulders he carried it to Cleonae. And finding Molorchus on the last of the thirty days about to sacrifice the victim to him as to a dead man, he sacrificed to Saviour Zeus and brought the lion to Mycenae. Amazed at his manhood, Eurystheus forbade him thenceforth to enter the city, but ordered him to exhibit the fruits of his labours before the gates. They say, too, that in his fear he had a bronze jar made for himself to hide in under the earth, and that he sent his commands for the labours through a herald, Copreus, son of Pelops the Elean. This Copreus had killed Iphitus and fled to Mycenae, where he was purified by Eurystheus and took up his abode.

1.1. The Tasks

The Nemean lion was ravaging the countryside near the town of Nemea, which is northwest of Mycenae. Hercules has three tasks to complete. After he tracks the lion, he shoots an arrow at the beast. First, he determines the speed at which his arrow strikes the invulnerable lion, given an angle of elevation and a distance to the lion (the **Shooting an Arrow** problem). In the second task, Hercules determines which set of regular polygons will allow him to tile the area of the cave mouth and trap the lion inside the cave (the **Closing the Cave Mouth** problem). As an associated exercise, Hercules chooses the mouth of the cave which will give him the greatest chance of finding the lion (the **Zeus Makes a Deal** problem).

1.1.1. Shooting an Arrow

TASK: Calculate the speed at which an arrow strikes the lion at a distance of 200 meters given a launch angle of 20°. Assume Hercules aims for the lion's head and shoulder area, which is the same distance off the ground as the arrow when it leaves Hercules' bow. Ignoring air resistance, how long does it take the arrow to travel from Hercules' bow to the lion?

1.1.2. Hercules Closes the Cave Mouth

TASK: To defeat the lion, Hercules must close up one cave entrance and attack the lion through the other. He finds several stacks of tiles nearby, each of which contains sets of regular polygons. There is one stack of equilateral triangles, one stack of squares, one stack of regular pentagons, one stack of regular hexagons, and one stack of regular octagons. Which stack(s) of polygons will allow Hercules to construct an edge-to-edge tiling in order to close up the mouth of the cave with no two tiles overlapping?

1.1.3. Exercise: Zeus Makes a Deal

TASK: Suppose that the cave has three, rather than two, mouths and that the lion is hiding just inside one of the mouths. Hercules selects one of the three cave mouths at random and is about to enter when Zeus, the king of the gods, suddenly tells him that the lion is not in a second cave mouth (not the one Hercules has choosen). Should Hercules change his mind and enter the remaining third mouth to the cave?

1.2. The Solutions

1.2.1. Shooting an Arrow

From Apollodorus:

First, Eurystheus ordered him to bring the skin of the Nemean lion; now that was an invulnerable beast begotten by Typhon. On his way to attack the lion he came to Cleonae and lodged at the house of a day-laborer, Molorchus; and when his host would have offered a victim in sacrifice, Hercules told him to wait for thirty days, and then, if he had returned safe from the hunt, to sacrifice to Saviour Zeus, but if he were dead, to sacrifice to him as to a hero. And having come to Nemea and tracked the lion, he first shot an arrow at him.

TASK: Calculate the speed at which an arrow strikes the lion at a distance of 200 meters given a launch angle of 20°. Assume Hercules aims for the lion's head and shoulder area, which is the same distance off the ground as the arrow when it leaves Hercules' bow. Ignoring air resistance, how long does it take the arrow to travel from Hercules' bow to the lion?

SOLUTION: Let's place a coordinate axis system so that the point where the arrow leaves Hercules' bow is at the origin when time $t - 0$. It is natural to assume constant acceleration and let the positive y-axis be vertically upward. The constant acceleration, directed only downward,

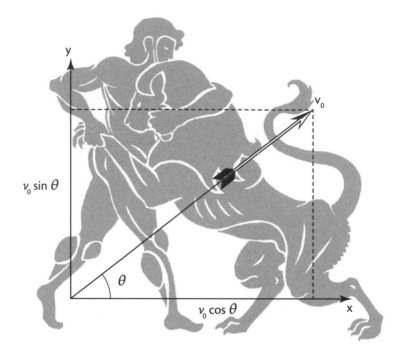

Figure 1.1. Determining the velocity components of Hercules' arrow.

is due to gravity and is denoted by $-g$. Further, think of this as a two-dimensional motion of the arrow through the air. We will neglect any effects the air might have on the arrow (in a simplified model, Hercules can neglect any effects of wind on the arrow's time of flight).

Because of our reference system, the initial position is given by $x(0) = y(0) = 0$. The initial velocity at time $t = 0$, which is at the exact instant the arrow begins its flight, is given by $v(0) = v_0$. With constant acceleration, we obtain the velocity by multiplying the acceleration by time t and adding the initial velocity:

$$v(t) = -gt + v_0. \tag{1.1}$$

Using trigonometry (see Figure 1.1), we can then determine the initial x- and y-components of v_0 as

$$v_{x0} = v_0 \cos \theta \quad \text{and} \quad v_{y0} = v_0 \sin \theta.$$

Since there is no acceleration in the x-direction, the horizontal component of velocity will remain constant. What does this mean? The horizontal velocity component ($v_0 \cos \theta$) will keep its initial value throughout the flight of the arrow. The vertical component, however, will change because of the downward acceleration. The x- and y-components of velocity (at any time t) become

$$v_x = v_0 \cos \theta \quad \text{and} \quad v_y = v_0 \sin \theta - gt.$$

The expression for v_y comes from replacing v_0 in Equation 1.1 with $v_0 \sin \theta$. Integrating these velocity components with respect to time, we can now determine the x- and y-components of the arrow's position at any time. These are given by

$$x = (v_0 \cos \theta)t \quad \text{and} \quad y = (v_0 \sin \theta)t - \frac{gt^2}{2}.$$

We can now calculate the horizontal distance (the range) that the arrow will travel before striking the lion. Setting the y-component to zero (recall that the vertical height where the arrow strikes the lion is the same as the height of the arrow leaving the bow), we obtain

$$(v_0 \sin \theta)t - \frac{gt^2}{2} = 0.$$

So, solving for t, we find that either $t = 0$ or

$$t = (2v_0 \sin \theta)/g. \tag{1.2}$$

Substituting this expression for t into the x-component, we find that the distance the arrow travels is

$$\text{distance} = (v_0 \cos \theta) \times \frac{2v_0 \sin \theta}{g}$$

$$= \frac{2v_0^2 \sin \theta \cos \theta}{g}.$$

We can now solve for the speed at which the arrow leaves Hercules' bow, v_0. Substituting in the launch angle ($\theta = 20°$), range (200 meters), and gravitational constant (9.8 meters per second squared), we find that

$$v_0^2 = \frac{200 \times 9.8}{2 \times \sin 20° \times \cos 20°}.$$

Evaluating and taking the square root of both sides,

$$v_0 \approx 55.22 \text{ meters/second} \approx 198.79 \text{ kilometers/hour.}$$

Therefore, Hercules must fire the arrow with an initial velocity of almost 200 kilometers per hour (125 miles per hour) from his mighty bow. This assumes that the speed of the arrow remains constant in flight (as mentioned above).

How long does it take the arrow to reach the lion after it leaves the bow? Substituting $v_0 \approx 55.22$ meters per second into Equation 1.2, we find

$$t = \frac{2v_0 \sin \theta}{g} = \frac{2 \times 55.22 \times \sin 20°}{9.8} \approx 3.85 \text{ seconds.}$$

Unfortunately, Hercules discovers that "the beast [is] invulnerable," as Apollodorus writes. Hercules' arrow will not penetrate the lion's hide. To capture the mighty lion, Hercules moves to another task in this labor.

1.2.2. Hercules Closes the Cave Mouth

From Apollodorus:

> And when the lion took refuge in a cave with two mouths, Hercules built up the one entrance and came in upon the beast through the other, and putting his arm round its neck held it tight till he had choked it; so laying it on his shoulders he carried it to Cleonae.

TASK: To defeat the lion, Hercules must close up one cave entrance and attack the lion through the other. He finds several stacks of tiles nearby, each of which contains sets of regular polygons. There is one stack of equilateral triangles, one stack of squares, one stack of regular pentagons, one stack of regular hexagons, and one stack of regular octagons (see Figure 1.2). Which stack(s) of polygons will allow Hercules to construct an edge-to-edge tiling in order to close up the mouth of the cave with no two tiles overlapping?

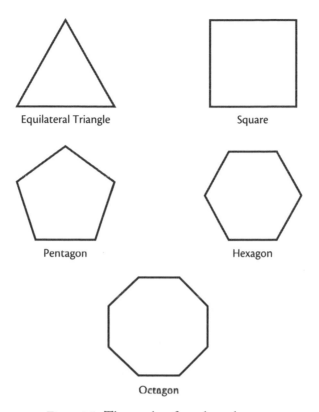

Figure 1.2. The stacks of regular polygons.

SOLUTION: Polygons are figures having many sides, usually four or more. A *tiling* is a covering of the entire plane with nonoverlapping figures. An *edge-to-edge tiling* is one in which the edge of a tile coincides completely with the edge of a bordering tile. We can assume that each entrance to the cave lies in a plane, so that the polygons can close it up by being stacked vertically edge upon edge (if possible). Hercules is not concerned with small gaps at the polygon/cave mouth interface.

Regular polygons were thought to have special meaning for the ancient Greeks, possibly beacuse of their simple yet high degree of symmetry. A *regular polygon* is a figure whose sides are equal and whose interior angles are also equal. Examples of regular polygons are equilateral triangles and squares. The ancient Greek philosopher Plato speculated that geometric solids formed from regular polygons (called *polyhedra*) were the shapes of the fundamental components

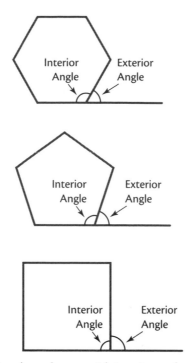

Figure 1.3. Regular polygons with interior and exterior angles.

of the physical universe. The word "polyhedra" also comes from the Greek and means "many faces." There are five basic Platonic solids: the *tetrahedron*, formed with 4 equilateral triangles; the *cube*, formed with 6 squares; the *octahedron*, formed with 8 equilateral triangles; the *dodecahedron*, formed with 12 pentagons; and the *icosahedron*, formed with 20 equilateral triangles. For the universe to be in harmony, there should be a one-to-one correspondence between the five solids and the primordial elements—earth, fire, water, and air. Therefore, Plato dedicated the dodecahedron (with its pentagonal faces) to the heavens and the rest to the planet Earth.

When Hercules takes a particular tile from one of the stacks, he measures the angles of the polygon. An *interior* angle is an angle formed between two adjacent sides of a polygon; an *exterior* angle is formed between one side and an extension of another side of a polygon. Figure 1.3 shows a few regular polygons with both interior and exterior angles depicted.

So, back to the task at hand, determining how many of each polygon can meet at a point. In order to have an edge-to-edge tiling, Hercules must fit the polygons together in such a way that their interior angles add up to 360° without overlap. Of the five regular polygons that are available to Hercules (triangle, square, pentagon, hexagon, and octagon), which may be used as an edge-to-edge tiling to block the mouth to the cave, thus trapping the Nemean lion inside? Each exterior angle in an n-sided polygon has $360/n$ degrees. As the number of sides of the polygon increases, the total of the interior angles also increases. In fact, the total of the interior angles is equal to $(n - 2) \times 180°$. The table below lists the five regular polygons and their respective interior and exterior angles.

Polygon	Interior Angle	Exterior Angle
Triangle	60°	120°
Square	90°	90°
Pentagon	108°	72°
Hexagon	120°	60°
Octagon	135°	45°

So, for any regular polygon having more than six sides, each interior angle must be greater than 120° and each exterior angle must be less than 60°. Notice that the octagon satisfies this requirement. Before we develop the solutions, photocopy Figure 1.2 and try to make a tiling with the regular polygons.

Let's start with the triangles. Six equilateral triangles come together at a single point. The sum of the interior angles is thus $6 \times 60° = 360°$. Hercules can use equilateral triangles to close the cave without overlap. Since the sum of the interior angles in a square is $4 \times 90° = 360°$, Hercules finds that squares will also suffice to close up the cave using an edge-to-edge tiling. What about the pentagon? If three pentagons come together, the sum of their interior angles will be $3 \times 108° = 324°$, so three pentagons would be too few. If four pentagons come together, the sum of their interior angles will be $4 \times 108° = 432°$, so four pentagons would be too many. Pentagons will not work.

Hercules then adds up the interior angles for hexagons and octagons. If three hexagons come together, the sum of their interior angles will be $3 \times 120° = 360°$, which shows that three hexagons will work as an edge-to-edge tiling. If three octagons come together, the sum of their interior angles will be $3 \times 135° = 405°$, so three octagons is too many. But if only two octagons are placed together, the sum of their interior angles will be $2 \times 135° = 270°$, which is too few.

Is the hexagon the polygon with the largest number of sides that will tile the plane? For what values of n does $180(n - 2)$ divide $360n$? Unless $2n/(n - 2)$ is an integer, an n-gon will not work. For $n \geq 7$,

$$2 < \frac{2n}{n - 2} < 3,$$

so $n = 6$ is the largest n-gon. Hercules can use the equilateral triangles, the squares, or the regular hexagons to close up one mouth of the cave. Then he can enter through the other mouth and trap and capture the lion.

1.2.3. Exercise: Zeus Makes a Deal

From Apollodorus:

And when the lion took refuge in a cave with two mouths, Hercules built up the one entrance and came in upon the beast through the other, and putting his arm round its neck held it tight till he had choked it.

TASK: Suppose that the cave has three, rather than two, mouths and that the lion is hiding just inside one of the mouths. Hercules selects one of the three cave mouths at random and is about to enter when Zeus, the king of the gods, suddenly tells him that the lion is not in a second cave mouth (not the one Hercules has choosen). Should Hercules change his mind and enter the remaining third mouth to the cave?

SOLUTION: First, if one of the cave mouths is eliminated, then there seems to be a 50-50 chance that the lion is in either of the two remaining cave mouths. Let's see if this is true.

As an example, assume that Hercules chooses cave mouth 1 and that Zeus has told him that the lion is *not* in cave mouth 2. Hercules must decide whether to switch his choice and enter cave mouth 3 instead of staying with cave mouth 1. There are three outcomes associated with Hercules' choice:

- Find the lion by switching cave mouths
- Don't find the lion by switching cave mouths
- Find the lion by not switching cave mouths

Is it advantageous to Hercules to switch after hearing what Zeus has to say?

We assume that no matter which cave mouth Hercules chooses, Zeus does two things: first, he reveals a mouth that Hercules has not chosen, and second, he does not reveal the actual cave mouth where the lion is (or perhaps the other gods might protest). To solve this task we model the problem as a conditional probability situation. Given two events, how does the information that event 2 has occurred affect the probability assigned to event 1?

We must determine the probability that Hercules will find the lion by switching, given that Zeus tells him the lion is not in one of the mouths. Is this an easy problem to solve? If Hercules develops a strategy to never switch, he will find the lion only one-third of the time. We could think of this in terms of the probability of failure, which is two-thirds. If Hercules does switch, then the probability of finding the lion (success) is two-thirds. In other words, if Hercules switches after receiving information from Zeus, then there are now only two possible entrances to the cave which will lead to success (finding the lion)– the entrance Hercules initially chose and the one that Zeus does not mention.

Using conditional probability, we could derive the mathematics behind the probabilities. However, it makes the discussion much more complicated and really is not necessary (I'll leave it up to you). For example, the probability that Zeus tells Hercules the lion is *not* in cave mouth 2 *given* it is actually in cave mouth 3 is exactly two-thirds. It is therefore advantageous for Hercules to switch and enter a different cave mouth.

There are many internet web sites dedicated to this problem, which is sometimes referred to as "The Let's Make a Deal Problem" or "The Monty Hall Paradox," after the game show host. Several include educational and fun Java applets as well. Give them a try. Type "Java Applet Let's Make a Deal" into your search engine. Pick any one of the applets. As you repeatedly try the applet (say 30 times), you should see that the results converge to the true probability of two-thirds success by switching and one-third success by not switching.

Here's another option: try playing a game with a friend. Your friend will be Hercules, and you will be Zeus. Select three playing cards (one of which represents the lion) and place them face down on a table. Only you (as the mighty Zeus) know which card is the lion. Have your friend select a card but not turn it over. As Zeus, turn over another card which is not the lion. Keep track of your friend's outcomes. After switching every time, the cumulative probability of success will approach two-thirds.

At long last, Hercules successfully captures and then chokes the lion. In paintings and sculptures, Hercules is often seen with the skin of the lion draped over his shoulders. Diodorus writes, "Since he could cover his whole body with it because of its great size, he had in it a protection against the perils which were to follow." Euripides corroborates this story, writing that Hercules "put its skin upon his back, hiding his yellow hair in its fearful tawny gaping jaws."

CHAPTER  2

The Second Labor: The Lernean Hydra

From Apollodorus:

As a second labour he [Eurystheus] ordered him to kill the Lernean hydra. That creature, bred in the swamp of Lerna, used to go forth into the plain and ravage both the cattle and the country. Now the hydra had a huge body, with nine heads, eight mortal, but the middle one immortal. So mounting a chariot driven by Iolaus, he came to Lerna, and having halted his horses, he discovered the hydra on a hill beside the springs of the Amymone, where was its den. By pelting it with fiery shafts he forced it to come out, and in the act of doing so he seized and held it fast. But the hydra wound itself about one of his feet and clung to him. Nor could he effect anything by smashing its heads with his club, for as fast as one head was smashed there grew up two. A huge crab also came to the help of the hydra by biting his foot. So he killed it, and in his turn called for help on Iolaus who, by setting fire to a piece of the neighboring wood and burning the roots of the heads with the brands, prevented them from sprouting. Having thus got the better of the sprouting heads, he chopped off the immortal head, and buried it, and put a heavy rock on it, beside the road that leads through Lerna to Elaeus. But the body of the hydra he slit up and dipped his arrows in the gall. However, Eurystheus said that this labour should not be reckoned among the ten because he had not got the better of the hydra by himself, but with the help of Iolaus.

2.1. The Tasks

According to myth, the hydra lived near the swamp of Lerna, which is southeast of Mycenae. The hydra ravaged livestock in the fields.

13

Its blood was poisonous and, by some accounts, even its breath could be deadly. Diodorus wrote that the hydra had 100 heads, but we shall consider the hydra with only 9 heads described by Apollodorus. The mighty Hercules has two tasks to complete this labor. First, he must deal with the hydra growing two new heads in the place where one is smashed. Let's model a situation where once Hercules seizes the hydra, he smashes two mortal heads every minute with his club (the **One Head Replaced by Two** problem). Then, in the second task, Hercules and Iolaus must cauterize the roots of the mortal heads to prevent them from sprouting new heads (the **Cauterizing the Hydra** problem).

2.1.1. One Head Replaced by Two

TASK: Once Hercules seizes the hydra, he smashes two mortal heads every minute with his club. However, where each head is smashed, two new heads grow. Determine the number of heads on the hydra after 15 minutes, after 30 minutes, and at any future time t. Further, determine if the number of heads on the hydra reaches an equilibrium value. By continuing to smash the hydra's heads, can Hercules "get ahead" and reach a point where the number of heads stays at a constant value? Assume that Iolaus has not yet cauterized any of the roots.

2.1.2. Cauterizing the Hydra

TASK: Hercules and Iolaus must stop the hydra's heads from multiplying. Let's assume that Iolaus attempts to set his burning brand onto the roots of the heads of the hydra with a 65 percent probability of success, as the hydra's heads (and headless necks) are swinging about wildly trying to fight Hercules. If he is unsuccessful, Iolaus tries again, up to four times in a minute. After a minute, it is too late to cauterize the neck, as two new heads suddenly grow in place of the one that Hercules lops off. What is the probability that he will *not* be successful in four tries? Finally, if Iolaus has to cauterize a total of 69 necks, what is the expected number of *misses* (i.e., how many new heads will be expected to grow)?

2.2. The Solutions

2.2.1. One Head Replaced by Two

From Apollodorus:

As a second labour he ordered him to kill the Lernean hydra. That creature, bred in the swamp of Lerna, used to go forth into the plain and ravage both the cattle and the country. Now the hydra had a huge body, with nine heads, eight mortal, but the middle one immortal. So mounting a chariot driven by Iolaus, he came to Lerna, and having halted his horses, he discovered the hydra on a hill beside the springs of the Amymone, where was its den. By pelting it with fiery shafts he forced it to come out, and in the act of doing so he seized and held it fast. But the hydra wound itself about one of his feet and clung to him. Nor could he effect anything by smashing its heads with his club, for as fast as one head was smashed there grew up two.

TASK: Once Hercules seizes the hydra, he smashes two mortal heads every minute with his club. However, where each head is smashed, two new heads grow. Determine the number of heads on the hydra after 15 minutes, after 30 minutes, and at any future time t. Further, determine if the number of heads on the hydra reaches an equilibrium value. By continuing to smash the hydra's heads, can Hercules "get ahead" and reach a point where the number of heads stays at a constant value? Assume that Iolaus has not yet cauterized any of the roots.

SOLUTION: Let $h(t)$ be the number of heads on the hydra at time t, measured in minutes. Initially, there are 9 heads, so $h(0) = 9$. We are given that only the mortal heads multiply after Hercules smashes them. We are told that Hercules smashes 2 mortal heads each minute. For instance, after 1 minute, Hercules has smashed 2 mortal heads, but 4 have grown back in their place, leaving a total of 11 heads. So, $h(1) = 11$. What about after 2 minutes? In that second minute, Hercules again smashes 2 more heads, but 4 grow back. In the first 2 minutes, Hercules has smashed 4 heads, but twice that number, or 8 heads, have grown back. So, $h(2) = 13$.

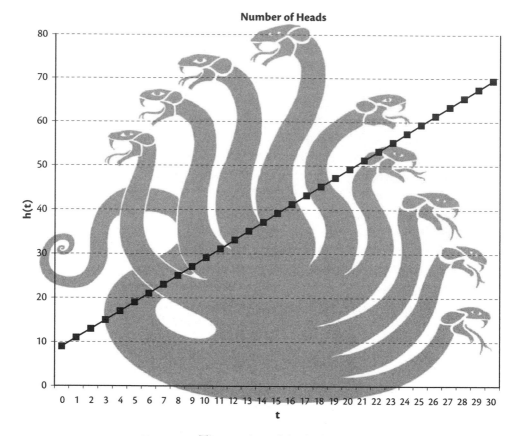

Figure 2.1. The number of the hydra's heads.

Think of the function $h(t)$ as an arithmetic sequence. At any future time k, $h(k) = 2k + h(0)$, or

$$h(k) = 2k + 9. \tag{2.1}$$

All we need to do now is replace k with any future time value to determine the number of heads. When $k = 15$, $h(15) = 2 \times 15 + 9 = 39$ heads. Further, $h(30) = 69$, and so forth. A plot of the number of heads with respect to time (Equation 2.1) is shown in Figure 2.1. As expected, it looks like a line.

Hercules must quickly deal with the increasing number of heads, determining how to stop them from multiplying. Does the number of heads reach some limit or equilibrium value? Determining whether or not equilibrium values exist and classifying them as stable or unstable immensely assists the process of analyzing the long-term behavior of a system. Remember that Equation 2.1 represents a line. This means that there is no equilibrium value. As k increases, so does $h(k)$. Incidentally, if Hercules were to work twice as fast, lopping off twice as many heads in a minute, he would simply be creating a more unwieldy hydra, one with many more heads than if he continued at his two-heads-per-minute pace. The number of heads will continue to increase unless Hercules somehow stops them from growing. That is the next task.

2.2.2. Cauterizing the Hydra

From Apollodorus:

> The hydra wound itself about one of Hercules' feet and clung to him. Nor could he effect anything by smashing its heads with his club, for as fast as one head was smashed there grew up two. In his turn he called for help on Iolaus who, by setting fire to a piece of the neighboring wood and burning the roots of the heads with the brands, prevented them from sprouting.

TASK: Hercules and Iolaus must stop the hydra's heads from multiplying. Let's assume that Iolaus attempts to set his burning brand onto the roots of the heads of the hydra with a 65 percent probability of success, as the hydra's heads (and headless necks) are swinging about wildly trying to fight Hercules. If he is unsuccessful, Iolaus tries again, up to four times in a minute. After a minute, it is too late to cauterize the neck, as two new heads suddenly grow in place of the one that Hercules lops off. What is the probability that he will *not* be successful in four tries? Finally, if Iolaus has to cauterize a total of 69 necks, what is the expected number of *misses* (i.e., how many new heads will be expected to grow)?

SOLUTION: Iolaus was Hercules' nephew, and he comes to Hercules' rescue by assisting him in burning the roots of the hydra's many heads. Each time Iolaus touches a burning brand to a mortal neck of the hydra, cauterization takes place. He continues to set his burning brand onto a neck of the hydra until he is successful in sealing it, which means that no new head can grow from that neck or until the minute expires and two new heads sprout. Iolaus repeats his trials until he achieves his first success or until two new heads appear.

Let n be the number of the trial on which Iolaus first succeeds. In this manner, n is not a fixed number but could be any number 1, 2, 3, and so forth. If p is the probability of success, then $(1 - p)$ is the probability of failure. Each attempt at sealing a neck is called a binomial trial. Since p is the same for each trial that Iolaus attempts, then the probability of success on the nth trial is given by

$$P(n) = p\,(1 - p)^{n-1}. \qquad (2.2)$$

We can assume the trials are independent and performed under identical conditions. Also, there are only two outcomes: success (p) and failure $(1 - p)$. In our problem, the probability of success is $p = 0.65$. Therefore, the probability of success on the first try is

$$P(1) = p(1 - p)^{1-1} = p = 0.65.$$

As for additional attempts, $P(2) = (0.65)(0.35)^1 = 0.2275$, or 22.75 percent. $P(3) = (0.65)(0.35)^2 = 0.0796$, or 7.96 percent.

We now use $P(1)$, $P(2)$, $P(3)$, and $P(4)$ ($= 0.0279 = 2.79$ percent). Since the trials for $n = 1$ or 2 or 3 or 4 are all mutually exclusive, we calculate the probability that any one event can occur by adding up the probabilities as

$$\begin{aligned} P(n = 1 \text{ or } 2 \text{ or } 3 \text{ or } 4) &= P(1) + P(2) + P(3) + P(4) \\ &= 0.65 + 0.2275 + 0.0796 + 0.0279 \\ &= 0.9850. \end{aligned}$$

The probability that Iolaus will be successful on a given neck before two new heads sprout is 98.5 percent. Not too bad. In fact, that's good news for Hercules, who must wrestle and defeat all of the heads.

What is the probability that Iolaus will not be successful in four tries? Since Iolaus has a probability of 98.5 percent of successfully cauterizing a neck of the hydra, the probability that he will not be successful is $1 - 0.985 = 0.015$, or 1.5 percent. We will need this value to answer the last question. If Iolaus must cauterize a total of 69 hydra roots of the heads, we would then expect $69 \times 0.015 = 1.04$, or just one hydra neck to sprout two more heads which Hercules must then smash with his club.

This task uses the geometric probability distribution, which is a discrete distribution related to the binomial distribution. It is used when the problem is concerned with obtaining the *first* success. Iolaus repeats his binomial ("go" or "no-go") trials of cauterizing the necks until he succeeds, and then he stops after achieving that success. From the first task of this labor, it took 30 minutes for the hydra to have 69 heads. With their high chance of success in cauterizing the necks, Hercules and Iolaus can be confident that almost all the heads will be sealed within $\frac{1}{2}$ hour. The sooner Iolaus joins in to singe the necks, the sooner Hercules will complete the task.

When the last mortal neck has been cauterized, Hercules chops off the immortal head and buries it, placing a heavy rock on top of the grave. He dips his arrows into the hydra's poisonous blood to use in later labors.

The Third Labor: The Hind of Ceryneia

From Apollodorus:

As a third labour he ordered him to bring the Cerynitian hind alive to Mycenae. Now the hind was at Oenoe; it had golden horns and was sacred to Artemis; so wishing neither to kill nor wound it, Hercules hunted it a whole year. But when, weary with the chase, the beast took refuge on the mountain called Artemisius, and thence passed to the river Ladon, Hercules shot it just as it was about to cross the stream, and catching it put it on his shoulders and hastened through Arcadia. But Artemis with Apollo met him, and would have wrested the hind from him, and rebuked him for attempting to kill her sacred animal. Howbeit, by pleading necessity and laying the blame on Eurystheus, he appeased the anger of the goddess and carried the beast alive to Mycenae.

3.1. The Tasks

Ceryneia is a remote mountain located in the northern part of the Peleponnesus peninsula. It lies about 80 kilometers northwest of Mycenae. Many legends exist about the famous Cerynetian hind, a sacred deer. Most sources agree that, although it had golden horns (antlers), the hind was indeed female. Hercules has two tasks to complete. First, he must determine where the deer of Ceryneia will cross the river Ladon in order to get from Mount Artemisius to a forest in Arcadia on the far side of the river in the shortest possible time (the **Optimizing the Hind's Journey** problem). Then Hercules must carry the hind from Mount Artemisius to Mycenae. In doing so, he calculates the amount of work needed to transport the Cerynitian hind along that great distance (the **Cerynitian Work** problem). As an associated exercise, suppose the force carried by Hercules is changing with each kilometer it is carried. For example, he decides to carry water for the journey, which he drinks as he takes the long trip to Eurystheus. What would be the new amount of work required (the **Work with a Variable Force** problem)?

3.1.1. Optimizing the Hind's Journey

TASK: According to legend, the Cerynitian deer was the fastest of deer. Assume that it can swim at a speed of 5 meters per second (taking the current into account) and that it can run on land even faster, at a speed of 8 meters per second over relatively long distances. The Ladon River is approximately 250 meters across from bank to bank. If a perpendicular line is drawn from Mount Artemisius across the river to a point B, the distance from B to a forest in Arcadia where it will seek shelter is 1600 meters. Assuming the Hind of Ceryneia will enter the river from Mount Artemisius, swim to the far bank, and then run along the shore, determine the landing point that will allow it to get from the mountain to the forest on the far side of the river in the shortest possible time.

3.1.2. Cerynitian Work

TASK: Once Hercules captures the hind at Artemisius, he must carry it to Eurystheus. Determine the work needed for Hercules to transport the Cerynitian hind from Mount Artemisius to Mycenae, which is a distance of 80 kilometers. Assume the hind has a mass of 125 kilograms.

3.1.3. Exercise: Work with a Variable Force

TASK: Suppose the force carried by Hercules is changing with each kilometer it is carried. For example, he decides to carry water for the journey, which he drinks as the trip to Mycenae progresses. He then sweats out the same amount, thus not gaining any weight himself.

If Hercules initially carries 40 kilograms of water, which is diminished by $\frac{1}{2}$ kilogram per kilometer, how much total work is being done?

3.2. The Solutions

3.2.1. *Optimizing the Hind's Journey*

From Apollodorus:

> As a third labour he ordered him to bring the Cerynitian hind alive to Mycenae. Now the hind was at Oenoe; it had golden horns and was sacred to Artemis; so wishing neither to kill nor wound it, Hercules hunted it a whole year. But when, weary with the chase, the beast took refuge on the mountain called Artemisius, and thence passed to the river Ladon, Hercules shot it just as it was about to cross the stream.

TASK: According to legend, the Cerynitian deer was the fastest of deer. Assume that it can swim at a speed of 5 meters per second (taking the current into account) and that it can run on land even faster, at a speed of 8 meters per second over relatively long distances. The Ladon River is approximately 250 meters across from bank to bank. If a perpendicular line is drawn from Mount Artemisius across the river to a point B, the distance from B to a forest in Arcadia where it will seek shelter is 1600 meters. Assuming the Hind of Ceryneia will enter the river from Mount Artemisius, swim to the far bank, and then run along the shore, determine the landing point that will allow it to get from the mountain to the forest on the far side of the river in the shortest possible time.

SOLUTION: A picture will help us best (see Figure 3.1). Let A be the deer's starting point on the Mount Artemisius side of the Ladon River and let F be the point downstream on the opposite shore of the Ladon corresponding to the location of the forest in Arcadia. Let B be the point directly across the river from Artemisius and let P be the point from which the sacred deer will swim to the opposite shore (P lies between B and F). As we assume in the problem, the deer's swimming speed accounts for any effect of the current of the river.

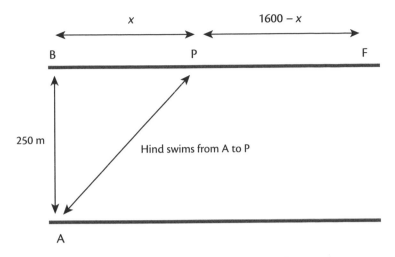

Figure 3.1. The Ladon Valley and the hind's journey.

If we let x be the distance from B to P, then the distance the deer will run is $|PF| = 1600 - x$ meters. Using the Pythagorean Theorem, Hercules determines the distance that the deer must swim as $|AP| = \sqrt{x^2 + (250)^2}$. As an aside, Pythagorus lived in the sixth century BC. Apollodorus, who wrote *The Library*, lived in the second century BC, so the Pythagorean Theorem was known to scholars at the time of his writings.

We wish to optimize the time, and knowing the rate, we have

$$\text{time} = \frac{\text{distance}}{\text{rate}}.$$

This allows us to solve for the time it takes the great deer to swim across the Ladon River and the time it takes to run from point P to the forest (point F). The swimming time is $\sqrt{x^2 + (250)^2}/5$ seconds, and the running time is $(1600 - x)/8$ seconds, so the total time T as a function of the distance x is

$$T(x) = \frac{\sqrt{x^2 + (250)^2}}{5} + \frac{1600 - x}{8}.$$

The distance x can take on values from 0 to 1600, so the domain of $T(x)$ is $[0, 1600]$. If $x = 0$, the deer swims from A to B (and then runs from B to F), and if $x = 1600$, the deer swims directly from A to F (no running involved).

To optimize the time (get from A to F in the shortest time), we need to find the critical point, x_{crit}, where the slope of the tangent line is zero. We can do this by taking the derivative of $T(x)$ which is

$$\frac{dT(x)}{dx} = \frac{x}{5\sqrt{x^2 + (250)^2}} - \frac{1}{8}.$$

Setting $\frac{dT(x)}{dx} = 0$, we find that

$$\frac{x}{5\sqrt{x^2 + (250)^2}} = \frac{1}{8}.$$

Cross-multiplying yields

$$8x = 5\sqrt{x^2 + (250)^2}.$$

Squaring both sides and solving for x yields

$$x^2 = \frac{(5)^2 \times (250)^2}{(8)^2 - (5)^2},$$

$$x_{crit} \approx \pm\, 200.16 \text{ meters.}$$

Notice that we get two solutions. Obviously, the negative solution is extraneous and not physically relevant. However, it is important to check for extraneous solutions in optimization problems.

How can we be sure that swimming from point A to a point 200.16 meters from point B is optimal? Can we call x_{crit} a global minimum? There are two methods of finding the absolute minimum of a continuous function on a closed interval. The only candidates for the global minimum must occur at either the end points or the critical points. Second, we could use the Second Derivative Test on the critical

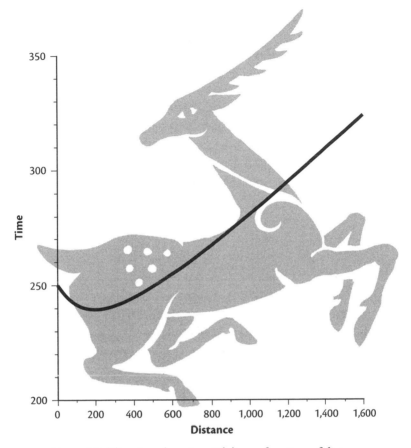

Figure 3.2. The time function $T(x)$ as a function of distance.

point in the interval. Only one of these methods is needed to find an absolute minimum.

If we zoomed in on the interval surrounding the local minimum in Figure 3.2, we would see that there is indeed a minimum value for the time when the distance is 200.16 meters. Using our first method, we find the time associated with the critical point ($x \approx 200.16$): $T(200.16) = 239.03$ seconds. Again, look at Figure 3.2. What about the end points of the interval? When $x = 0$, $T(0) = 250.00$ seconds, and when $x = 1600$, $T(1600) \approx 323.88$ seconds. Both of the function's values at the end points are greater than that at the critical point. Since $T(x)$ is a continuous function, this tells us we have a minimum at x_{crit}.

Alternatively, we can apply the Second Derivative Test. Differentiating $T(x)$ twice with respect to x, we find

$$\frac{d^2 T(x)}{dx^2} = -\frac{x^2}{5\left[x^2 + (250)^2\right]^{3/2}} + \frac{1}{5\sqrt{x^2 + (250)^2}}.$$

Evaluating the second derivative at the critical point yields $T''(x_{\text{crit}}) \approx$ 0.00038. Although very small, this is still a positive value. Since the second derivative is positive, the time function is concave up at the critical point. Having $T'' > 0$ tells us that we have a minimum when $x_{\text{crit}} \approx 200.16$. Further, it is a global minimum because there are no other critical values.

So what? The deer must swim from Mount Artemisius on the near shore (point A) to a point about 200 meters from B on the far shore and then run $1600 - 200 = 1400$ meters to point F and the safety of the Arcadian forest. The total time required to make the trip is just under 4 minutes. What does all of this mean for Hercules? Since the great Hind of Ceryneia is so much faster running on land than swimming in the river, it swims to a point close to Artemisius, only 200 meters downstream and across the river, affording Hercules the opportunity to shoot and capture it before it can use its great speed on land.

3.2.2. Cerynitian Work

From Apollodorus:

> As a third labour he ordered him to bring the Cerynitian hind alive to Mycenae. Now the hind was at Oenoe: catching it Hercules put it on his shoulders and hastened through Arcadia.

TASK: Once Hercules captures the hind at Artemisius, he must carry it to Eurystheus. Determine the work needed for Hercules to transport the Cerynitian hind from Mount Artemisius to Mycenae, which is a distance of 80 kilometers. Assume the hind has a mass of 125 kilograms.

SOLUTION: Simply put, *work* is the amount of effort required to perform a given task. In a physics sense, work depends on the amount of force involved. In the situation where acceleration is constant, work can be thought of as the quantity required to move a force through a distance. If the force is constant (nonvarying),

$$\text{work} = \text{force} \times \text{distance}.$$

It is about 80 kilometers from Cerynea to Mycenae. That's 80,000 meters. If we assume that Hercules carries the beast over one shoulder and that the height above the ground is constant for the entire trip (he may switch shoulders, but he doesn't carry it on his hip, for example), then the amount of work needed is equal the animal's force, or weight, times the 80,000-meter distance. The force of the hind is equal to its mass times the acceleration due to gravity (in units of newtons). Thus,

$$\text{work} = 125 \text{ kilograms} \times 9.8 \ \frac{\text{meters}}{\text{seconds}^2} \times 80,000 \text{ meters}$$

$$= 98,000,000 \text{ newton-meters}.$$

It will take over 98 million newton-meters (or joules) of work to transport the Cerynitian hind to Mycenae. This does not include the amount of work needed to hoist the deer to Hercules' shoulder (about 2 meters off the ground) each time he lifts the animal, assuming he does not carry the deer in one straight trip. Each time Hercules lifts the hind off the ground, he does an additional $125 \times 9.8 \times 2 = 2450$ joules of work. To put this amount of work into perspective, 1 joule is approximately the work required to raise a small object weighing 102 grams 1 meter (moving against the gravity of the earth) or the energy required to lift a baseball (145 grams) 70 centimeters off the ground.

3.2.3. Exercise: Work with a Variable Force

From Apollodorus:

As a third labour he ordered him to bring the Cerynitian hind alive to Mycenae. Now the hind was at Oenoe: catching it Hercules put it on his shoulders and hastened through Arcadia.

TASK: Suppose the force carried by Hercules is changing with each kilometer it is carried. For example, he decides to carry water for the journey, which he drinks as the trip to Mycenae progresses. He then sweats out the same amount, thus not gaining any weight himself. If Hercules initially carries 40 kilograms of water, which is diminished by $\frac{1}{2}$ kilogram per kilometer, how much total work is being done?

SOLUTION: We now need to use calculus to determine the work. The work can be found by evaluating the definite integral of the variable force function using the distance as the limits of integration. We must subtract $\frac{1}{2}$ kilogram for every kilometer (or every 1000 meters) traveled. The hind and water initially have a combined mass of $125 + 40 = 165$ kilograms and will have a mass of 125 kilograms when Hercules reaches Mycenae. Let us define the force function as $F(x)$, where

$$F(x) = \left[165 - \left(\frac{1}{2} \times \frac{x}{1000}\right)\right] \times 9.8$$

$$= 1617 - \frac{49x}{10000}. \tag{3.1}$$

The distance x is measured in meters ($x = 0$ corresponds to Mount Artemisius in Ceryneia, and $x = 80000$ corresponds to Mycenae). Note that $F(80000) = 1617 - \frac{49 \times 80000}{10000} = 1225$ newtons, which is the same force as in the second task (since the water is all gone). The work $W(x)$ can be found by evaluating

$$W(x) = \int_0^{80000} \left(1617 - \frac{49x}{10000}\right) dx$$

$$= 1617x - \frac{4.9x^2}{20000}\Big|_0^{80000}$$

$$= 113,680,000 \text{ joules.}$$

Transporting the great deer and the extra water causes Hercules to exert 113.68 million joules of work, with about 15.68 million extra joules due just to the weight of the water.

The Fourth Labor: The Erymanthian Boar

From Apollodorus:

As a fourth labour he ordered him to bring the Erymanthian boar alive; now that animal ravaged Psophis, sallying from a mountain which they call Erymanthus. So passing through Pholoe he was entertained by the centaur Pholus, a son of Silenus by a Melian nymph. He set roast meat before Hercules, while he himself ate his meat raw. When Hercules called for wine, he said he feared to open the jar which belonged to the centaurs in common. But Hercules, bidding him be of good courage, opened it, and not long afterwards, scenting the smell, the centaurs arrived at the cave of Pholus, armed with rocks and firs. The first who dared to enter, Anchius and Agrius, were repelled by Hercules with a shower of brands, and the rest of them he shot and pursued as far as Malea. Thence they took refuge with Chiron, who, driven by the Lapiths from Mount Pelion, took up his abode at Malea. As the centaurs cowered about Chiron, Hercules shot an arrow at them, which, passing through the arm of Elatus, stuck in the knee of Chiron. Distressed at this, Hercules ran up to him, drew out the shaft, and applied a medicine which Chiron gave him. But the hurt proving incurable, Chiron retired to the cave and there he wished to die, but he could not, for he was immortal. However, Prometheus offered himself to Zeus to be immortal in his stead, and so Chiron died. The rest of the centaurs fled in different directions, and some came to Mount Malea, and Eurytion to Pholoe, and Nessus to the river Evenus. The rest of them Poseidon received at Eleusis and hid them in a mountain. But Pholus, drawing the arrow from a corpse, wondered that so little a thing could kill such big fellows; howbeit, it slipped from his hand and lighting on his foot killed him on the spot. So when Hercules returned to Pholoe, he beheld Pholus dead; and he buried

him and proceeded to the boar hunt. And when he had chased the boar with
shouts from a certain thicket, he drove the exhausted animal into deep snow,
trapped it, and brought it to Mycenac.

4.1. The Tasks

The Erymanthian boar lived on Mount Erymanthus in Arcadia in
the central Peleponnesus. To accomplish this labor, Hercules has
three tasks to complete. First, after Hercules calls for wine, he must
help the centaurs decide a wine distribution issue (**The Centaurs'
Wine** exercise). When Hercules accidentally shoots Chiron with the
poisonous arrow, he must determine how long it will take for the
centaur to die (the **Chiron's Poison** problem). For the third and final
task, Hercules chases the boar into deep, unpacked snow, which causes
the animal's energy level to decrease. We model the energy of the boar
using a half-life decay scenario (**The Capture of the Boar** problem).
According to some sources, Hercules used a large net to trap and
capture the boar in the snow.

4.1.1. Exercise: The Centaurs' Wine

TASK: Hercules calls for wine, and Pholus poses the following problem.
Five centaurs have 45 jars of wine, of which 9 each are full, three-
quarters full, half-full, one-quarter full, and empty. The centaurs want
to divide the wine and the jars without transferring the wine from jar
to jar in such a way that each receives the same amount of wine and the
same number of jars, and further so that each receives at least one of
each kind of jar and no two of them receive the same number of every
kind of jar. Can the wine jars be so divided?

4.1.2. Chiron's Poison

TASK: As the centaurs are hiding around Chiron, Hercules shoots an
arrow at them, which unfortunately hits Chiron in the knee. Distressed,
Hercules runs up to his friend and draws out the shaft, but the poison
has entered Chiron's bloodstream. The centaur has an immune system

which will combat the effects of the poison. Some poisons have specific antidotes which either prevent the poison from working or reverse the effects of the poison. Because of his expert medical knowledge, Chiron tells Hercules to administer a certain medicine, which works as an antidote by bolstering the centaur's natural detoxification abilities. The poison breaks down 25 percent of Chiron's immune system in 5-minutes. Assume that Chiron's immune system protection level takes on values from 0 to 1. To combat the effects of the poison, Hercules administers medicine to Chiron every 5 minutes, which improves Chiron's protection level by adding a value of 0.1 to it. When Chiron's immune system protection falls below the 0.5 level, the poison causes the centaur great pain. How long does it take until Chiron is constantly in great pain?

4.1.3. The Capture of the Boar

TASK: Hercules is intent on completing his assigned labor, so he must force the boar to become cold and tired. The boar enters the deep, unpacked snow running at 3 meters per second. The wintry conditions cause the boar's energy to decrease at a rate proportional to the amount of energy it has remaining, which in turn causes its speed to decrease accordingly. When running in unpacked snow, it takes 47 minutes for the boar to lose half of its energy. When the energy level of the boar decreases to 15 percent of the energy it started with, it is considered exhausted. Assuming the boar starts running in the deep snow at full energy, how long must Hercules chase the boar in the deep snow?

4.2. The Solutions

From Apollodorus:

As a fourth labour he ordered him to bring the Erymanthian boar alive; now that animal ravaged Psophis, sallying from a mountain which they call Erymanthus. So passing through Pholoe he was entertained by the centaur Pholus, a son of Silenus by a Melian nymph. He set roast meat before Hercules, while he himself ate his meat raw. When Hercules called for wine, he said he feared to open the jar which belonged to the centaurs in common.

4.2.1. Exercise: The Centaurs' Wine

TASK: Hercules calls for wine, and Pholus poses the following problem. Five centaurs have 45 jars of wine, of which 9 each are full, three-quarters full, half-full, one-quarter full, and empty. The centaurs want to divide the wine and the jars without transferring the wine from jar to jar in such a way that each receives the same amount of wine and the same number of jars, and further so that each receives at least one of each kind of jar and no two of them receive the same number of every kind of jar. Can the wine jars be so divided?

SOLUTION: This is a variation of an ancient problem, from *The Greek Anthology*, proposed during the Middle Ages. It has been adjusted slightly to fit this situation. First, let's state the obvious. Each centaur must have nine ($45 \div 5 = 9$) jars of wine, each with varying amounts (including none) of wine. Since the jars of wine are all filled with multiples of one-quarter of a jar (including zero), each centaur must have

$$\frac{1}{5} \times [9 + (9 \times 2) + (9 \times 3) + (9 \times 4)] = 18$$

quarter-jars of wine. We will designate the respective numbers of each type of wine which any one centaur might receive as A, B, C, D, and E, where A is a full jar, B is a three-quarters-full jar, and so forth. This yields the following two equations:

$$4A + 3B + 2C + D = 18, \qquad (4.1)$$

$$A + B + C + D + E = 9. \qquad (4.2)$$

Equation 4.1 is the amount of wine (in quarter-jars), and Equation 4.2 tells us how many of each volume type makes up the total of 9 jars per centaur.

Since the total amount of wine from the 9 jars must sum to 18 quarter-jars, at most, only one centaur can have 3 full jars of wine (convince yourself why). If he were to get 4 full jars, for example, then having one each of the three-quarters-full, half-full, and quarter-full jars would give that centaur a total of 22 quarter-jars—too much! That's leads us to the conclusion that the largest number of full jars that any centaur can have is 3.

We must use Equations 4.1 and 4.2 to solve the task. Excluding the solutions with zero values (remember, each centaur must get at least one of each volume), there are eight possible solutions to these equations (you should verify this):

Solution	A	B	C	D	E
I	3	1	1	1	3
II	2	1	2	3	1
III	2	1	3	1	2
IV	2	2	1	2	2
V	1	1	5	1	1
VI	1	2	3	2	1
VII	1	3	1	3	1
VIII	1	3	2	1	2

There are three possibilities where a centaur gets 2 full jars of wine, and there are four possibilities where a centaur gets only 1 full jar of wine.

Let's examine one of the solutions. Solution I means that one centaur gets 3 full jars of wine, 1 three-quarters-full jar, 1 half-full jar, 1 quarter-full jar, and 3 empty jars. Numerically, this looks like

$$3 + \left(1 \times \frac{3}{4}\right) + \left(1 \times \frac{1}{2}\right) + \left(1 \times \frac{1}{4}\right) + (3 \times 0)$$
$$= 18 \text{ quarter-jars of wine.}$$

Any one of these eight solutions would satisfy any one of the centaurs in the stated problem. However, for the five centaurs we need five solutions, one for each. Furthermore, we only have 9 full jars of wine ($9 \times A$), 9 three-quarters-full jars ($9 \times B$), and so forth. The problem now becomes one in which the centaurs choose their assignments in such a way that all of the As add up to 9, all of the Bs sum to 9, with the same conditions for C, D, and E.

Let's discuss solution V in some detail. If one of the centaurs receives 5 jars filled halfway with wine, there would be no way to give the other four centaurs each a single half-full jar. Why? Only solutions I, IV, and VII offer 1 half-full jar each (look at column C); by using solution V, we would be forced to use solutions I, IV, VII, and either II or VIII. Then the total number of half-full jars would be $5 + 1 + 1 + 1 + 2 = 10 \neq 9$. So, cross out solution V; we cannot use it. After eliminating solution V, the sums of columns A through E become 12, 13, 13, 13, and 12, respectively. Therefore, to keep the volume sum at 9, we will have to omit two solutions for which the sums in these columns are 3, 4, 4, 4, and 3.

Solution I cannot be eliminated; it alone contains a 3 in the first and last columns; plus it provides only 1 jar in the three-quarters-full, half-full, and quarter-full categories. However, solutions II and VIII, solutions III and VII, and solutions IV and VI give the needed sums, so we now have three fundamental sets of five solutions. Each of these solutions gives $5! = 120$ permutations among the five centaurs. The fundamental sets are solutions I, III, IV, VI, and VII; solutions I, II, IV, VI, and VIII; and solutions I, II, III, VII, and VIII. Any one of these three sets of solutions will satisfy the task.

4.2.2. Chiron's Poison

From Apollodorus:

> As the centaurs cowered about Chiron, Hercules shot an arrow at them, which, passing through the arm of Elatus, stuck in the knee of Chiron. Distressed at this, Hercules ran up to him, drew out the shaft, and applied a medicine which Chiron gave him. But the hurt proving incurable, Chiron retired to the cave and there he wished to die, but he could not, for he was immortal. However, Prometheus offered himself to Zeus to be immortal in his stead, and so Chiron died.

TASK: As the centaurs are hiding around Chiron, Hercules shoots an arrow at them, which unfortunately hits Chiron in the knee. Distressed, Hercules runs up to his friend and draws out the shaft, but the poison has entered Chiron's bloodstream. The centaur has an immune system which will combat the effects of the poison. Some poisons have specific antidotes which either prevent the poison from working or reverse the effects of the poison. Because of his expert medical knowledge, Chiron tells Hercules to administer a certain medicine, which works as an antidote by bolstering the centaur's natural detoxification abilities. The poison breaks down 25 percent of Chiron's immune system in 5-minutes. Assume that Chiron's immune system protection level takes on values from 0 to 1. To combat the effects of the poison, Hercules administers medicine to Chiron every 5 minutes, which improves Chiron's protection level by adding a value of 0.1 to it. When Chiron's immune system protection falls below the 0.5 level, the poison causes the centaur great pain. How long does it take until Chiron is constantly in great pain?

SOLUTION: In Greek mythology, Chiron was regarded above all other the centaurs. Unlike the other centaurs, who were notorious for being overly indulgent drinkers and party animals (literally), Chiron was intelligent, civilized, and kind. He was known for his knowledge and skill with medicine, and several Greek authors mention that he tutored the great Greek heroes Achilles, Jason, and even Hercules.

Hercules removes the arrow from Chiron's knee, but the damage has already been done. Recall that the arrow had been dipped in the gall from the body of the hydra (the Second Labor), so the tip is fatally poisonous. One drop of the poison from Hercules' arrow will instantly kill a mortal, but Chiron is immortal. The poison spreads quickly, breaking down the protection provided by Chiron's immune system to 75 percent of its previous level in 5-minutes. So, every 5 minutes, Hercules applies the medicine prescribed by Chiron. Our task is to model the amount of protection from the immune system and determine when it stays below 50 percent, meaning that further doses of the medicine will no longer help the doomed centaur.

Let $p(t)$ be the protection offered by Chiron's immune system at any time t, which we measure in 5-minute discrete intervals. Initially, Chiron's protection is 100 percent. From one 5-minute time period to the next, Chiron's immune system protection drops 25 percent. However, Hercules administers medicine that bolsters the immune system protection level by adding a value of 0.1 to it. Mathematically, we can describe this system as

$$p(t+1) = 0.75\,p(t) + 0.1. \tag{4.3}$$

Additionally, $p(0) = 1.00$ (his immune system offers total protection before the poison goes to work). Let's iterate a few steps to see what happens to $p(t)$ after 5 or 10 minutes. Beginning with $p(0)$, we find that

$$p(1) = 0.75\,p(0) + 0.1 = (0.75 \times 1) + 0.1 = 0.85,$$

$$p(2) = 0.75\,p(1) + 0.1 = (0.75 \times 0.85) + 0.1 = 0.7375.$$

We could continue to iterate in this fashion, and it appears that the protection level will continue to decrease. Does it reach an equilibrium level? The solution has the form

$$p(k) = c(0.75)^k + d,$$

where c and d are constants and k is some future time value. Substituting t and $t+1$ for k, we have

$$p(t) = c(0.75)^t + d,$$

$$p(t+1) = c(0.75)^{t+1} + d.$$

We now substitute these two expressions into Equation 4.3:

$$c(0.75)^{t+1} + d = 0.75[c(0.75)^t + d] + 0.1.$$

Solving for d, we find $d = 0.4$. Thus, the general solution becomes

$$p(k) = c(0.75)^k + 0.4.$$

Implementing the initial condition ($p(0) = 1$) gives $c = 0.6$, so the particular solution for the amount of protection provided by Chiron's immune system at any time k is

$$p(k) = 0.6(0.75)^k + 0.4. \tag{4.4}$$

Let's check this out. After 5 minutes, $p(1) = 0.6\,(0.75)^1 + 0.4 = 0.85$ and $p(2) = 0.7375$. We plot the particular solution in Figure 4.1. Notice that after 20 minutes (when $k = 4$), the protection level has

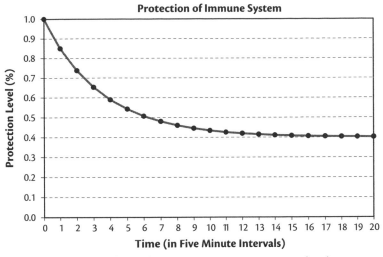

Figure 4.1. Chiron's immune system protection level.

fallen below 60 percent. Solving Equation 4.4 for k with $p(k) = 0.5$ yields

$$k = \frac{\ln(1/6)}{\ln(3/4)} \approx 6.2283$$

time periods (just past 31 minutes). Chiron will continue to be in great pain after this time.

Regarding an equilibrium level, $p(t)$ will approach the equilibrium level of 40 percent after about 1 hour and stay there until Chiron dies. Ironically, Chiron, the master of the healing arts, could not heal himself, so he willingly gave up his immortality, choosing to trade his life for the release of Prometheus. As a reward, Chiron was placed in the sky as the constellation Sagittarius (also known as Centaurus). Grieving for his two friends, Chiron and Pholus, Hercules is both saddened and angered. This fills him with adrenaline, and he now plans his capture of the Erymanthian boar.

4.2.3. The Capture of the Boar

From Apollodorus:

[Hercules] proceeded to the boar hunt. And when he had chased the boar with shouts from a certain thicket, he drove the exhausted animal into deep snow, trapped it, and brought it to Mycenae.

TASK: Hercules is intent on completing his assigned labor, so he must
force the boar to become cold and tired. The boar enters the deep,
unpacked snow running at 3 meters per second. The wintry conditions
cause the boar's energy to decrease at a rate proportional to the amount
of energy it has remaining, which in turn causes its speed to decrease
accordingly. When running in unpacked snow, it takes 47 minutes for
the boar to lose half of its energy. When the energy level of the boar
decreases to 15 percent of the energy it started with, it is considered
exhausted. Assuming the boar starts running in the deep snow at full
energy, how long must Hercules chase the boar in the deep snow?

SOLUTION: This is a *half-life* problem, as we are given information
about how long it takes for the boar to lose half of its energy. Let $b(t)$
be the energy level of the boar at any time t (measured in minutes).
Since the boar's energy decreases at a rate proportional to the amount
of energy it has remaining, we can write

$$\frac{db}{dt} = -kb, \tag{4.5}$$

where k is an unknown constant of proportionality. The negative sign
before k indicates that the energy is decreasing. We can solve this using
separation of variables (an exercise left for the reader), yielding

$$b(t) = b_0 e^{-kt}. \tag{4.6}$$

The term b_0 is the initial energy level of the boar at time $t = 0$. This
solution indeed satisfies Equation 4.5 (differentiate both sides, and
you should obtain the differential equation). We know that it takes
47 minutes for the energy level to reach $b_0/2$, so we substitute this
information into Equation 4.6 and obtain

$$\frac{b_0}{2} = b_0 e^{-k \times 47},$$

or

$$\frac{1}{2} = e^{-47k}.$$

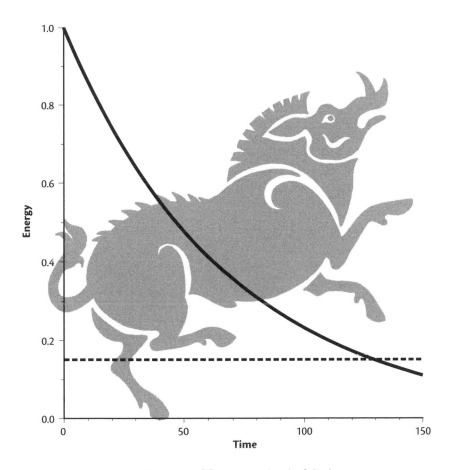

Figure 4.2. The energy level of the boar.

Taking the natural logarithm of both sides yields

$$k = -\frac{\ln\left(\frac{1}{2}\right)}{47}.$$

Substituting this back into Equation 4.6 shows that

$$b(t) = b_0 \left(\frac{1}{2}\right)^{t/47}. \qquad (4.7)$$

All that is left now is to determine how long it takes to get to 15 percent of b_0. Substitute $0.15\,b_0$ into the left-hand side of Equation 4.7

				A	N		H	
T							N	E
			E	H		Y		
		I					Y	
Y			T	N	H			R
	E					H		
		E		T	I			
M	Y							T
	I		Y	R				

Figure 4.3. Erymanthian sudoku puzzle.

and solve for t. We find that

$$t = 47 \times \frac{\ln(0.15)}{\ln(0.50)} \approx 128.64 \text{ minutes.}$$

It will take just over 2 hours for Hercules to chase and capture the exhausted Erymanthian boar. A plot of the solution (Equation 4.7) with $b_0 = 1$ is shown in Figure 4.2. When the solution curve crosses the line where the energy equals 0.15, the boar is exhausted.

Hercules captures the boar (probably with his net) and carries it back to Eurystheus in Mycenae.

4.3. The Erymanthian Sudoku Puzzle

Four labors down, 8 to go. You have now solved one-third of the 12 labors. As a reward, we offer some recreational mathematics in the form of a sudoku puzzle. In Figure 4.3, you'll see the popular puzzle with a twist: the numbers have been replaced by letters. Good luck! In the realm of sudoku difficulty levels (easy, medium, hard, and forget about it), this puzzle is classified as *hard*. After all, none of Hercules' labors were *easy*.

The Fifth Labor: The Augean Stables

From Apollodorus:

The fifth labour he laid on him was to carry out the dung of the cattle of Augeas in a single day. Now Augeas was king of Elis; some say that he was a son of the Sun, others that he was a son of Poseidon, and others that he was a son of Phorbas; and he had many herds of cattle. Hercules accosted him, and without revealing the command of Eurystheus, said that he would carry out the dung in one day, if Augeas would give him the tithe of the cattle. Augeas was incredulous, but promised. Having taken Augeas's son Phyleus to witness, Hercules made a breach in the foundations of the cattle-yard, and then, diverting the courses of the Alpheus and Peneus, which flowed near each other, he turned them into the yard, having first made an outlet for the water through another opening. When Augeas learned that this had been accomplished at the command of Eurystheus, he would not pay the reward; nay more, he denied that he had promised to pay it, and on that point he professed himself ready to submit to arbitration. The arbitrators having taken their seats, Phyleus was called by Hercules and bore witness against his father, affirming that he had agreed to give him a reward. In a rage Augeas, before the voting took place, ordered both Phyleus and Hercules to pack out of Elis. So Phyleus went to Dulichium and dwelt there, and Hercules repaired to Dexamenus at Olenus. He found Dexamenus on the point of betrothing perforce his daughter Mnesimache to the centaur Eurytion, and being called upon by him for help, he slew Eurytion when that centaur came to fetch his bride. But Eurystheus would not admit this labour either among the ten, alleging that it had been performed for hire.

5.1. The Tasks

Augeas reigned in the district of Elis in the northwestern Peleponnesus. His father had given him many herds of cattle, but after years of neglect in cleaning up after the cattle, the stables were over a meter deep in dung. Our hero has three tasks to complete. We start with a simple algebraic puzzle to determine how many herds of cattle Augeas owned (**The Herds of Augeas** problem). Next, suppose that Hercules had not "made breaches in the foundation of the cattle-yard." An aerial view of Olympia in Elis, where Augeas ruled, shows a valley with an elevated plateau. The stables that contain the cattle sit at the edge of this plateau, and they measure 750 meters long, 400 meters wide, and 2 meters high. The two rivers flow on either side of the stables, toward the edge of the plateau. Hercules must determine the hydrostatic pressure that the water will exert on the stable walls (the **Hydrostatic Pressure on the Stable Walls** problem). Finally, as Hercules diverts the Alpheus and Peneus rivers, he creates a mixing problem involving water and cattle dung. He must determine know how long it will take to rid the stables of the dreaded cattle dung (the **Cleaning the Stables with Torricelli** problem).

5.1.1. The Herds of Augeas

TASK: The following task is taken verbatim from *The Greek Anthology*, an ancient collection of epigrams:

> Hercules the mighty was questioning Augeas, seeking to learn the number of his herds, and Augeas replied: "About the streams of Alpheius, my friend, are half of them; the eighth part pasture around the hill of Cronos, the twelfth part far away by the precinct of Taraxippus; the twentieth part feed in holy Elis, and I left the thirtieth part in Arcadia; but here you see the remaining fifty herds."

How many herds did Augeas have?

5.1.2. Exercise: Hydrostatic Pressure on the Stable Walls

TASK: In the cattle yard at Elis, where only some of the total number of cattle are situated, Augeas had built stables in the shape of a rectangle 750 meters long, 400 meters wide, and 2 meters high. Calculate the volume of water needed to fill the stables to a height of 2 meters. What is the pressure at the base of the stables compared to the pressure $\frac{1}{2}$ meter off the ground? Further, what is the total force at each end of the rectangular stables (where the stables are 750 meters wide)? The density of water is 1000 kilograms per cubic meter.

5.1.3. Cleaning the Stables with Torricelli

TASK: Let's assume that Hercules must fill the stables with water to the full height of 2 meters to ensure that all of the dung is dissolved and then washed away. Model the situation with the two rivers (the Alpheus and the Peneus—see Figure 5.2) flowing into the valley. Suppose that the flow of each river into the stables is at a rate of 30 and 25 cubic meters per second, respectively. Suppose also that Hercules digs a hole at the opposite end that is a 2-meter-diameter circle underneath the stables. The amount of dung in the yard is initially 25,000 kilograms. Note: An average cow produces 14 to 18 kilograms of dung per day, but we'll assume that all of the cattle are not always inside the stables. Further assume that the stables fill up with water before Hercules opens the hole. Determine the total time required to cleanse the stables, meaning that no dung or water is left inside the cattle yard. Can the stables be cleaned in one day?

5.2. The Solutions

5.2.1. The Herds of Augeas

From Apollodorus:

The fifth labour he laid on him was to carry out the dung of the cattle of Augeas in a single day. Now Augeas was king of Elis; some say that he was a son of the Sun, others that he was a son of Poseidon, and others that he was a son of Phorbas; and he had many herds of cattle. Hercules accosted him, and without revealing the command of Eurystheus, said that he would carry out the dung in one day, if Augeas would give him the tithe of the cattle.

TASK: The following task is taken verbatim from *The Greek Anthology*, an ancient collection of epigrams:

> Hercules the mighty was questioning Augeas, seeking to learn the number of his herds, and Augeas replied: "About the streams of Alpheius, my friend, are half of them; the eighth part pasture around the hill of Cronos, the twelfth part far away by the precinct of Taraxippus; the twentieth part feed in holy Elis, and I left the thirtieth part in Arcadia; but here you see the remaining fifty herds."

How many herds did Augeas have?

SOLUTION: This particular riddle comes from Book XIV of *The Greek Anthology*, which contains arithmetical problems, riddles, and oracles. Classics scholar W. R. Paton wrote that the arithmetical problems in Book XIV of *The Greek Anthology* were written down by Metrodorus, who was probably a grammarian living in the time of Constantine the Great (306 to 337). There are exactly 150 such problems and puzzles, and Paton claims they "can be easily solved by algebra." Our task is number 4 of 150 and is entitled, "On the Dung of Augeas."

Now to the easy algebra. Let's define n as the number of herds of cattle that Augeas owned. From the epigram, we set up the following equation:

$$n = \frac{n}{2} + \frac{n}{8} + \frac{n}{12} + \frac{n}{20} + \frac{n}{30} + 50.$$

Multiplying by the common denominator, 120, gives $120n = 95n + 6000$. So $25n = 6000$, or $n = 240$ herds. That implies a lot of dung for Hercules to clean up!

5.2.2. Exercise: Hydrostatic Pressure on the Stable Walls

From Apollodorus:

> The fifth labour he laid on him was to carry out the dung of the cattle of Augeas in a single day. Now Augeas was king of Elis; some say that he was a son of the Sun, others that he was a son of Poseidon, and others that he was a son of Phorbas; and he had many herds of cattle.

TASK: In the cattle yard at Elis, where only some of the total number of cattle are situated, Augeas had built stables in the shape of a rectangle 750 meters long, 400 meters wide, and 2 meters high. Calculate the volume of water needed to fill the stables to a height of 2 meters. What is the pressure at the base of the stable compared to the pressure $\frac{1}{2}$ meter off the ground? Further, what is the total force at each end of the rectangular stable (where the stables are 750 meters wide)? The density of water is 1000 kilograms per cubic meter.

SOLUTION: First, we should calculate the volume of water needed to fill the stables to a height of 2 meters. The volume of the rectangular pen is

$$\text{volume} = \text{area} \times \text{height}.$$

With the dimensions given above, we obtain a volume of

$$\text{volume} = 750 \times 400 \times 2 = 600,000 \text{ meters}^3.$$

To fill the trough to the top of the stable walls, Hercules needs to divert 600,000 cubic meters of water! This amounts to approximately 158.5 million gallons of water. By comparison, the Lincoln Memorial Reflecting Pool between the Lincoln Memorial and the Washington Monument in Washington, D.C., is approximately 618 meters long, 51 meters wide, and about $\frac{1}{2}$ meter deep, giving it a volume of just under 16,000 cubic meters (or 6.75 million gallons). We assume that all of the dung dissolves in the water so that it can be washed away as the water flows out. This phenomenon is unlikely, but this assumption will simplify the problem.

Consider the example of submerging a thin horizontal plate in a fluid. The plate has an area A (measured in square meters), and the fluid has a density of δ kilograms per cubic meter at a depth of y meters below the water's surface. The fluid that is directly above the horizontal plate has a volume which we calculate as

$$\text{volume} = \text{area} \times \text{depth}.$$

To calculate the pressure, set the origin to be at the surface of the water, with the downward direction considered positive. Thus, the depth of the water is 2 meters, and we need to sum up the pressure times the area of a cross section of the rectangular pen. Modeling this hydrostatic pressure problem, we should realize that the water pressure increases with depth (it becomes greater as we get closer to the bottom of the stables). Why? The weight of the water increases with the distance from the surface.

So, the volume $V = A \times y$. Further, knowing the density and the volume, we can calculate the mass of the fluid as the product of the mass density and the volume, or $m = \delta V = \delta A y$. The force F that the fluid exerts on the thin horizontal plate is therefore

$$F = mg = \delta A y g,$$

where g is the acceleration due to gravity, acting in a downward direction (hence positive, in this case).

Pressure P is defined as a force per unit area provided the pressure is constant over the specified area. Dividing the force by the area, we find that $P = \delta g y$. An essential principle in physics concerning the pressure of a fluid is that *at any point in a fluid, the pressure is the same in all directions*. (For example, if Hercules dives into a pond, he feels the same pressure at all points on his body at a given depth.)

The depth of the water in the Augean stables is 2 meters. Since we are given the density of water as $\delta = 1000$ kilograms per cubic meter, the water pressure at ground level is

$$P = \delta g d = 1000 \times 9.8 \times 2 = 19,600 \; \frac{\text{newtons}}{\text{meter}^2}.$$

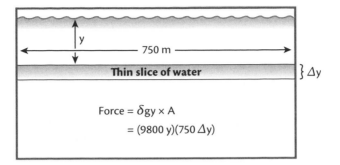

Figure 5.1. Calculating the force on the wall.

At $\frac{1}{2}$ meter above the ground, the pressure is

$$P = 1000 \times 9.8 \times 1.5 = 14,700 \; \frac{\text{newtons}}{\text{meter}^2}.$$

Since the pressure varies linearly with depth ($P = \delta g y$), we expect the pressure three-quarters of the way from the surface to be three-quarters of that at the bottom.

To calculate the force on the walls, we divide the stables into thin horizontal strips of height Δy (assuming the pressure at the depth of each strip is constant—see Figure 5.1). Looking at the stables from the end where the width is 400 meters, the area of each strip is $750 \times \Delta y$ square meters. At a depth of y meters, the pressure is $\delta g y = 9800 y$ newtons per square meter. Thus, the force is

$$F = \delta y g \times A = (9800 y) * (750 \, \Delta y) = (7.35 \times 10^{6}) \; y \, \Delta y.$$

We sum each of the strips at depths ranging from 0 to 2 meters and take the limit as $\Delta y \to 0$, obtaining

$$\text{Total force} = \int_{0}^{2} (7.35 \times 10^{6}) \, y \, dy$$

$$= (3.675 \times 10^{6}) \, y^{2} \Big|_{0}^{2}$$

$$= 1.47 \times 10^{7} \text{ newtons.} \tag{5.1}$$

Hopefully the walls of the stables can sustain the force.

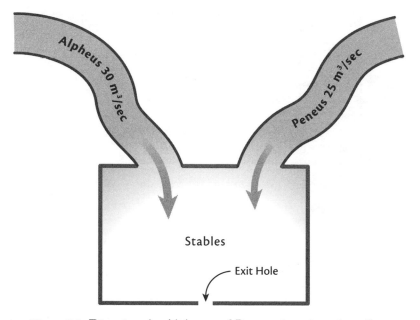

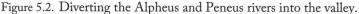

Figure 5.2. Diverting the Alpheus and Peneus rivers into the valley.

5.2.3. Cleaning the Stables with Torricelli

From Apollodorus:

Hercules accosted him [Augeas], and without revealing the command of
Eurystheus, said that he would carry out the dung in one day, if Augeas
would give him the tithe of the cattle. Augeas was incredulous, but promised.
Having taken Augeas's son Phyleus to witness, Hercules made a breach in
the foundations of the cattle-yard, and then, diverting the courses of the
Alpheus and Peneus, which flowed near each other, he turned them into the
yard, having first made an outlet for the water through another opening.

TASK: Let's assume that Hercules must fill the stables with water to
the full height of 2 meters to ensure that all of the dung is dissolved and
then washed away. Model the situation with the two rivers (the Alpheus
and the Peneus—see Figure 5.2) flowing into the valley. Suppose that
the flow of each river into the stables is at a rate of 30 and 25 cubic
meters per second, respectively. Suppose also that Hercules digs a hole
at the opposite end that is a 2-meter-diameter circle underneath the

stables. The amount of dung in the yard is initially 25,000 kilograms. Note: An average cow produces 14 to 18 kilograms of dung per day, but we'll assume that all of the cattle are not always inside the stables. Further assume that the stables fill up with water before Hercules opens the hole. Determine the total time required to cleanse the stables, meaning that no dung or water is left inside the cattle yard. Can the stables be cleaned in one day?

SOLUTION: Think of this problem as pulling the stopper from an old-fashioned bath tub filled with water. How long will it take to fill up and then drain? Does the water drain at a uniform rate? Before finding out how long it takes to drain, let's determine the time needed to fill the stables with water from the two rivers. We are given the input flow rates which total 55 cubic meters per second. (As a point of reference, the Mississippi River unleashes about 17,000 cubic meters per second as it passes New Orleans, and the Hudson River flows at about 110 cubic meters per second as it nears New York City.) To fill the stables (600,000 cubic meters) would take 10,909 seconds, or 3.03 hours, assuming a constant rate of flow. A relationship called Torricelli's Law allows us to describe how quickly the water flows out once Hercules creates an outlet.

Evangelista Torricelli was an Italian scientist and mathematician who lived from 1608 to 1647. Torricelli was appointed to succeed Galileo upon Galileo's death in 1642 as the court mathematician to Grand Duke Ferdinando II of Tuscany, at the University of Florence. Despite a relatively short life—he died of typhoid fever at the age of 39—he made some significant discoveries. He was the first person to create a sustained vacuum. His proposal of an experiment demonstrating that atmospheric pressure determines the height to which a fluid rises in a tube inverted over the same liquid ultimately led to the development of the mercury barometer. In 1644, Torricelli published his proof that the velocity v of a fluid through a hole, under the influence of gravity alone, is proportional to the square root of the height of the fluid h above the hole. Mathematically, we write this as

$$v = \frac{dh}{dt} = -k\sqrt{2gh}, \qquad (5.2)$$

where g is the acceleration due to gravity and k is a constant of proportionality. Equation 5.2 is equivalent to the speed that the fluid would have in a free fall from a height h. To show that this is true, we know that volume V equals area times height. If the cross-sectional area A remains constant, then

$$\frac{dV}{dt} = A\frac{dh}{dt}.$$

Torricelli's Law tells us that

$$\frac{dV}{dt} = -a\sqrt{2gh}.$$

In this case, the constant of proportionality is the cross-sectional area a of the hole through which the fluid is draining. Equating these two volume rate expressions, we have

$$\frac{dh}{dt} = -\frac{a}{A}\sqrt{2gh}. \tag{5.3}$$

In our problem, $A = 750 \times 400 = 300,000$ meters^2 and $a = \pi(1)^2 = \pi$ meters^2. The initial condition is $h(0) = 2$ meters. Assuming the dung dissolves completely in the water which is washed away, the initial amount of cattle dung is not relevant. We can solve Equation 5.3 using separation of variables.

We keep the independent variable terms (t) and constants on the right-hand side and the dependent variable terms ($h(t)$) on the left-hand side, obtaining

$$\frac{dh}{\sqrt{h}} = -\frac{a\sqrt{2g}}{A}\,dt.$$

Integrating both sides, we get

$$2\sqrt{h} = -\frac{a\sqrt{2g}}{A}t + C, \text{ or}$$

$$h = \left(-\frac{a\sqrt{2g}}{2A}t + \frac{C}{2}\right)^2,$$

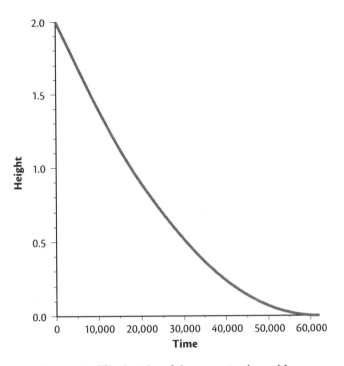

Figure 5.3. The height of the water in the stables.

where C is a constant of integration. Using the initial condition of $h(0) = 2$, we solve and obtain $C = 2\sqrt{2}$. Therefore (after substituting in values for a, A, and g), the height of the water in the stables at any time t is given by

$$h(t) = \left(-\frac{\sqrt{19.6}\,\pi}{600000}\,t + \sqrt{2}\right)^2. \qquad (5.4)$$

A plot of Equation 5.4 is shown in Figure 5.3. You can see that the rate of flow out of the hole at the bottom of the stables starts off quickly and then slows as the height goes to zero. This is intuitive, as we would not expect the flow rate out of the stables to be uniform. Solving for the time when the height is zero, we find that $t = 61008.2$ seconds (we can ignore the negative solution as extraneous), or 16.95 hours for the time it takes the stables to drain. That brings the total time required to fill and drain the stables to just under 20 hours. Hercules did it in one day!

A different approach to this problem would be to ask for the size of the exit hole that Hercules must create in order for the stables to drain in under 24 hours. That exercise is left to the reader. Incidentally, by today's standards, Hercules' method of washing away the dung by diverting the Alpheus and Peneus rivers would not be acceptable in the beef cattle industry.

You can try this experiment as well. See the Bibliography for an article by Farmer and Gass describing the use of Torricelli's Law in classroom applications (they use a 2-liter soda bottle with a small hole —having a diameter of 4 millimeters—near the bottom; they also stress having another container to catch the outflow—a good idea!). I have used a sun tea jar, which works nicely as long as you can measure the area of the opening at the bottom and the cross-sectional area of the jar (which should be constant). Fill your cylindrical container with a known quantity of fluid. Tape a ruler onto the jar or mark centimeter values on the container with a thin erasable marker. Add the water to the jar and get ready to experiment. Open the spigot and start timing the decline of the fluid's height. When you reach different time intervals (5 seconds, 10 seconds, 15 seconds, and so forth), record the height of the water. Perform the experiment a few times. Compare your experimental values to the values predicted by Torricelli's Law.

CHAPTER  6

The Sixth Labor: The Stymphalian Birds

From Apollodorus:

The sixth labour he enjoined on him was to chase away the Stymphalian birds. Now at the city of Stymphalus in Arcadia was the lake called Stymphalian, embosomed in a deep wood. To it countless birds had flocked for refuge, fearing to be preyed upon by the wolves. So when Hercules was at a loss how to drive the birds from the wood, Athena gave him brazen castanets, which she had received from Hephaestus. By clashing these on a certain mountain that overhung the lake, he scared the birds. They could not abide the sound, but fluttered up in a fright, and in that way Hercules shot them.

6.1. The Tasks

These birds swarmed in flocks around the Stymphalian Lake in Arcadia. By several accounts, the birds had arrow-firing wings and armor-piercing beaks. Hercules has three tasks to complete. For the first task, Hercules attempts to drive the birds out of the woods by running in an ever-increasing spiral, waving his club and yelling. We must determine how far Hercules runs (**The Spiral of Archimedes** problem). After an unsuccessful attempt at routing the birds, Hercules uses the bronze castanets given to him by Athena. He clashes the castanets on the side of the mountain, and the noise resonates into the woods. The second task forces Hercules to determine the frequency that will cause resonance (the **Resonating Castanets** problem). I caution you that this problem will require some knowledge of differential equations to solve. Then, as the birds take flight, Hercules must shoot them before they get too far from the woods. As an exercise, he numerically approximates the value of π (the **Monte Carlo Shooting Scheme** exercise) while shooting the birds.

6.1.1. The Spiral of Archimedes

TASK: Initially, Hercules hopes to drive the birds out of the woods by running in an ever-increasing spiral, waving his club and yelling. His path follows a simple curve, known as the *Spiral of Archimedes*, represented by the function $r = 50\theta$, given in polar coordinates, where r is the radius measured in meters and θ is measured in radians. If Hercules traverses two complete turns around the spiral, how far has he run (what is his *arc length*)? What happens to the arc length if Hercules runs one additional revolution?

6.1.2. Resonating Castanets

TASK: When the solution of a linear ordinary differential equation exhibits an oscillatory nature, sinusoidal driving forces may produce a peculiar behavior known as *resonance*. To drive the birds from the trees, Hercules must set up identical metal rods at the edge of the forest, to act as harmonic oscillating tuning forks, which have a natural circular frequency of 256 hertz. Using the bronze castanets from the goddess Athena, he bangs them together at a periodic rate (modeled by the forcing function $\cos(\omega t)$). What should the circular frequency of the castanets be in order to induce resonance in the metal rods, thus causing the birds to flee? The vibration caused by the banging must be loud enough to scare the birds into flight. How many metal rods must he put in proximity of the castanets in order to raise the amplitude of the castanets to a value of $\frac{1}{2}$ after 10 seconds? Assume the metal rods, modeled as a forced harmonic oscillator, are initially at rest. What happens in the case where $\omega = 80\pi$, close to but not exactly 256 hertz?

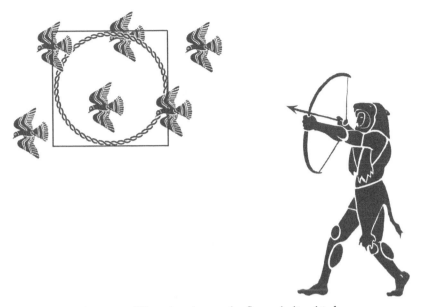

Figure 6.1. Hercules shoots the Stymphalian birds.

6.1.3. Exercise: Monte Carlo Shooting Scheme

TASK: We expect that after scaring the birds into taking flight, Hercules will shoot them down. Imagine that he has forced them to fly in a certain direction and that their flight path takes them in front of a large square marked in the trees by a rope. Inscribed inside the square is a circle. Hercules stands perpendicular to the bird's flight path and shoots his arrows at them as they cross in front of him (see Figure 6.1). Although this task is not explicitly described by Apollodorus, perhaps Hercules is trying to impress the gods by approximating π.

6.2. The Solutions

6.2.1. The Spiral of Archimedes

From Apollodorus:

The sixth labour he enjoined on him was to chase away the Stymphalian birds. Now at the city of Stymphalus in Arcadia was the lake called Stymphalian, embosomed in a deep wood. To it countless birds had flocked for refuge, fearing to be preyed upon by the wolves.

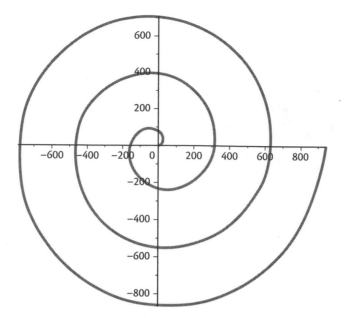

Figure 6.2. The Spiral of Archimedes.

TASK: Initially, Hercules hopes to drive the birds out of the woods by running in an ever-increasing spiral, waving his club and yelling. His path follows a simple curve, known as the *Spiral of Archimedes*, represented by the function $r = 50\theta$, given in polar coordinates, where r is the radius measured in meters and θ is measured in radians. If Hercules traverses two complete turns around the spiral, how far has he run (what is his *arc length*)? What happens to the arc length if Hercules runs one additional revolution?

SOLUTION: The *Spiral of Archimedes* is a simple curve denoted by the polar equation $r = a\theta$, where a is equal to 50 meters for Hercules. A plot of this function is shown in Figure 6.2.

The legends of Hercules predate Archimedes by about 400 years. Archimedes was probably born in Syracuse (a city on the island of Sicily) in the year 287 BC, and he died during the attack on Syracuse by the Romans in the Second Punic War (212 BC). Archimedes may have spent a considerable portion of his life in Alexandria, where he studied with the successors of Euclid. Although Archimedes achieved fame from his mechanical inventions, he believed that pure mathematics was the only worthy pursuit. He invented many devices, including the compound pulley, a long-range catapult, and the water screw, and he wrote about the principles of the fulcrum (seesaw). Several of his mechanical inventions were used by Syracuse against the Romans. When he contemplated a problem of moving a given weight by a given force, he supposedly uttered his famous saying, "Give me a place to stand on, and I can move the earth." Perhaps the most popular story involving Archimedes illustrates his intense preoccupation with mathematics and how he literally immersed himself in his work. He discovered, while taking a bath, a solution to the question of whether a certain crown supposed to be made of gold actually contained some silver, causing him to run naked through the streets to his house shouting, "Eureka!" Many historians believe that Archimedes died as he had lived, consumed by mathematical contemplation. The historian Plutarch provides the following account of Archimedes' demise:

> As fate would have it, he [Archimedes] was intent on working out some problem with a diagram and, having fixed his mind and his eyes alike on his investigation, he never noticed the incursion of the Romans nor the capture of the city. And when a soldier came up to him suddenly and bade him to follow, he refused to do so until he had worked out his problem to a demonstration; whereat the soldier was so enraged that he drew his sword and slew him.

Archimedes requested that a cylinder circumscribing a sphere be placed upon his tomb, together with an inscription giving the ratio of the cylinder to the sphere. He considered this ratio his greatest discovery.

Back to his famous spiral. First, we must calculate the arc length bounded by two complete turns, where the range of angle θ is $0 \le \theta \le 4\pi$. In Cartesian coordinates, if x and y are defined parametrically, the arc length formula is given by

$$ds = \sqrt{\left(\frac{dx}{dt}\right)^2 + \left(\frac{dy}{dt}\right)^2}. \qquad (6.1)$$

In polar coordinates, we convert x and y into polar coordinates using

$$x(t) = r(t) \cos \theta(t),$$

$$y(t) = r(t) \sin \theta(t).$$

To compute ds, we must first calculate the derivatives of $x(t)$ and $y(t)$ with respect to t. Using the product and chain rules,

$$\frac{dx}{dt} = (\cos \theta)\frac{dr}{dt} - (r \sin \theta)\frac{d\theta}{dt}$$

and

$$\frac{dy}{dt} = (\sin \theta)\frac{dr}{dt} + (r \cos \theta)\frac{d\theta}{dt}.$$

Substituting these two derivatives into Equation 6.1 allows us to give the arc-length formula in terms of the desired polar parametric coordinates. The new equation becomes

$$ds = \sqrt{\left(\frac{dr}{dt}\right)^2 + \left(r\frac{d\theta}{dt}\right)^2}\; dt. \qquad (6.2)$$

One more simplification is needed. In the case where the polar-coordinate equation is of the form $r = f(\theta)$, as in the Spiral of Archimedes, we may use the variable θ as a parameter, and the arc-length formula simplifies to

$$ds = \sqrt{\left(\frac{dr}{d\theta}\right)^2 + r^2}\; d\theta. \qquad (6.3)$$

We leave it as an exercise to show that Equation 6.2 is equivalent to Equation 6.3.

Knowing Equation 6.3, we can now calculate $dr/d\theta$ and determine the total distance traveled after two revolutions by integrating ds as θ ranges from 0 to 4π:

$$s = \int_0^{4\pi} \sqrt{\left(\frac{dr}{d\theta}\right)^2 + r^2}\; d\theta.$$

Substituting in $r = 50\theta$ and $\frac{dr}{d\theta} = 50$, we find

$$s = \int_0^{4\pi} \sqrt{(50)^2 + (50\theta)^2}\, d\theta$$

$$= 50 \int_0^{4\pi} \sqrt{1 + \theta^2}\, d\theta$$

$$= 50 \left[\frac{\theta\sqrt{1+\theta^2}}{2} + \frac{1}{2}\ln\left(\theta + \sqrt{1+\theta^2}\right) \right]_0^{4\pi}$$

$$\approx 4040.97 \text{ meters.}$$

Here we used the substitution $\theta = \tan x$ to evaluate this integral. Thus, Hercules travels just over 4 kilometers in running two revolutions of the Spiral of Archimedes. If one more revolution is added, Hercules must run (and we must integrate) from 0 to 6π, and the arc length increases to 8985.89 meters, or close to 9 kilometers!

Try plotting the Spiral of Archimedes with an online applet (type "Spiral of Archimedes Applet" into your internet search engine). Most applets offer animations and allow you to vary the parameter a, meaning that Hercules' actual position on the curve changes to correspond to the parametric values. Give it a try!

6.2.2. Resonating Castanets

From Apollodorus:

So when Hercules was at a loss how to drive the birds from the wood, Athena gave him brazen castanets, which she had received from Hephaestus. By clashing these on a certain mountain that overhung the lake, he scared the birds.

TASK: When the solution of a linear ordinary differential equation exhibits an oscillatory nature, sinusoidal driving forces may produce a peculiar behavior known as *resonance*. To drive the birds from the trees, Hercules must set up identical metal rods at the edge of the forest, to act as harmonic oscillating tuning forks, which have a natural circular frequency of 256 hertz. Using the bronze castanets from the goddess Athena, he bangs them together at a periodic rate (modeled by the forcing function $\cos(\omega t)$). What should the circular frequency of the castanets be in order to induce resonance in the metal rods, thus causing the birds to flee? The vibration caused by the banging must be loud enough to scare the birds into flight. How many metal rods must he put in proximity of the castanets in order to raise the amplitude of the castanets to a value of $\frac{1}{2}$ after 10 seconds? Assume the metal rods, modeled as a forced harmonic oscillator, are initially at rest. What happens in the case where $\omega = 80\pi$, close to but not exactly 256 hertz?

SOLUTION: Musical instruments and other objects are set into vibration at their natural frequencies when they are struck, strummed, or disturbed. This vibrational motion is caused by the input of energy at the object's natural frequency. The tendency of one object to force another adjoining or interconnected object into vibrational motion is often referred to as *forced vibration*. In musical applications, this increases the amplitude and therefore the loudness of the sound.

Consider a tuning fork. If you hold a tuning fork in your hand and strike it with a rubber mallet, a sound is produced. How? The tines of the tuning fork set surrounding air particles into vibrational motion producing sound. Further, if the tuning fork is set on another object, such as a glass, the glass begins to vibrate at the same natural frequency of the tuning fork. The tuning fork forces the surrounding glass particles to vibrate, which in turn forces surrounding air particles into vibrational motion, resulting in increased amplitude of the motion. This creates a louder sound.

Now consider a related situation. Suppose that a tuning fork is mounted in a fixed position and that a second tuning fork having the same natural frequency (say 256 hertz) is placed near the first fork. At first, neither of the tuning forks is vibrating. Then the first tuning fork is struck with a rubber mallet, and the tines begin to vibrate at its natural frequency of 256 hertz. Sound is produced. Then the tines of the tuning fork are grabbed to prevent their vibration, and remarkably the sound of 256 hertz is still heard. Only now the sound is being produced by the second tuning fork—the one which wasn't hit with the mallet. Wow! Try this in a classroom. What is happening?

In this demonstration, one tuning fork forces another tuning fork to vibrate at the same natural frequency. As the air particles surrounding the first fork (and its connected sound box) begin vibrating, the pressure waves it creates begin to impinge upon the second tuning fork at a periodic and regular rate of 256 hertz. Since the incoming sound waves have the same natural frequency as the second tuning fork, the second fork easily begins to vibrate at its natural frequency. This is an example of resonance—when one object vibrating at the same natural frequency of a second object forces the second object into vibrational motion.

In our situation, Hercules must know that one castanet vibrating at a certain frequency will cause a second object (the metal rod tuning fork) to vibrate if the natural frequency of the metal rod is the same or close to that of the castanet. Here is an experiment you can try: set a ping-pong ball on a stand so that it just contacts one of two tuning forks. Strike the other tuning fork and bring it close to the first. The ping-pong ball will start to move. This "agitation" should indicate the vibration of the tuning fork which is in contact with the ball.

Resonance occurs when two interconnected objects share the same vibrational frequency. When one of the objects is vibrating, it forces the second object into vibrational motion. The result is a large vibration, and if a sound wave within the audible range of human hearing is produced, a loud sound is heard. If the natural circular frequencies are close but not exactly the same, the phenomenon of a *beat* is created.

Back to the task. In this situation, there is nothing to smother the vibrating sound, so we have an *undamped harmonic oscillator* with a cosine forcing function. We model this with the following differential equation

$$y'' + \omega_0^2 y = \cos(\omega t), \tag{6.4}$$

where ω_0 and ω are both constants and $y(t)$ is the position of the oscillator (the metal rods) at some time t. The nonhomogeneous driving term ($\cos(\omega t)$) has circular frequency ω. The character of the solution depends greatly on whether or not ω equals ω_0. If it does, then resonance occurs. The system is responding to a bounded input with an unbounded output. We are given that $\omega_0 = 256$ hertz. In the first case, $\omega = 256$, matching ω_0.

The solution to Equation 6.4 has two parts, a solution to the associated homogeneous equation, labeled $y_h(t)$, and the solution to the nonhomogeneous equation, labeled $y_p(t)$. First we solve the homogeneous equation, $y'' + \omega_0^2 y = 0$. The initial conditions are that the system is at rest, so $y(0) = y'(0) = 0$. The differential equation's characteristic equation is

$$r^2 + \omega_0^2 = 0.$$

This yields $r = \sqrt{-\omega_0^2} = \pm\omega_0 i$. Therefore, using Euler's Formula, $y_h(t) = c_1 \cos(\omega_0 t) + c_2 \sin(\omega_0 t)$. We will solve for the unknown constants shortly.

To obtain the nonhomogeneous solution, since $\omega = \omega_0$, we conjecture

$$y_p(t) = At \cos(\omega_0 t) + Bt \sin(\omega_0 t).$$

We must multiply both terms by t since the forcing term matches the homogeneous solution. Differentiating twice gives

$$y_p''(t) = -(256)^2 At \ \cos(256t) - 512A \ \sin(256t)$$
$$-(256)^2 Bt \ \sin(256t) + 512B \ \cos(256t).$$

Substituting $y_p''(t)$ and $y_p(t)$ into Equation 6.4, with $\omega = \omega_0$, and solving for A and B delivers $A = 0$ and $B = \frac{1}{512}$. Thus,

$$y_p(t) = \frac{t}{512} \sin(\omega_0 t),$$

and the general solution is

$$y(t) = y_h(t) + y_p(t) = c_1 \cos(\omega_0 t) + c_2 \sin(\omega_0 t) + \frac{t}{512} \sin(\omega_0 t).$$

We now apply the initial conditions, $y(0) = y'(0) = 0$, which after substitution yields $c_1 = c_2 = 0$, and the solution to the driven

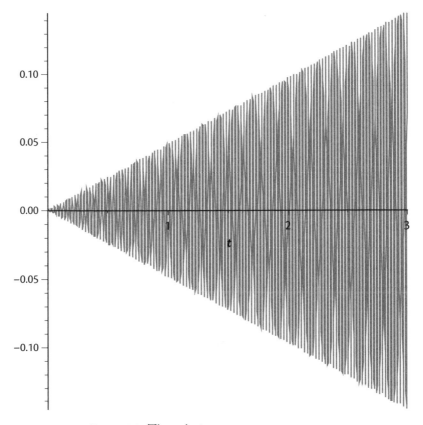

Figure 6.3. The solution to castanet resonance.

undamped harmonic oscillator (Equation 6.4) is

$$y(t) = \frac{t}{512} \sin(\omega_0 t). \tag{6.5}$$

If we plot $y(t)$ versus t, we notice that the amplitude continues to grow with t. It will grow without bound (see Figure 6.3). The sound will continue to get louder and louder.

To answer the task, how many rods will it take to raise the amplitude of the castanets to a value greater than $\frac{1}{2}$ after 10 seconds? We need to find k so that $ky(10) > \frac{1}{2}$. The ampliude's local maximum immediately before $t = 10$ can be found by solving $\frac{2\pi n}{256} = 10$ (from the argument of the forcing function), which gives $n = 407$. Therefore, on the 407th oscillation (corresponding to time $t = \frac{2\pi \times 407}{256} = 9.989$ seconds), $y(t)$ is at the local maximum immediately preceding $t = 10$. Since the sine function has a maximum value of 1, we have

$$\frac{9.989k}{512} \geq \frac{1}{2},$$

or $k \geq 26$. That means that Hercules must place 26 sets of metal rods within the resonating distance of the castanets. By this we mean that each rod will be forced into vibration by the others.

To answer the final question, when $\omega = 80\pi$, the differential equation becomes

$$y'' + \omega_0^2 y = \cos(80\pi t). \tag{6.6}$$

The solution to Equation 6.6 using the same initial conditions is

$$y(t) = \frac{25}{256} \frac{\cos(256t)}{25\pi^2 - 256} - \frac{25 \cos(80\pi t)}{6400\pi^2 - (256)^2}.$$

This solution is plotted in Figure 6.4. This phenomenon is called a *beat*, which exhibits a sinusoid of circular frequency with a slowly varying amplitude which is itself sinusoidal. The amplitude grows and diminishes. This is similar to the swelling and fading sounds made by a pair of tuning forks with similar, but not exactly the same, natural frequencies. The beat frequency is $(256 - 80\pi)/2$ hertz, and the pitch heard is $(256 + 80\pi)/2$ hertz.

Once Hercules uses the castanets to create a resonating sound, the birds take flight and our hero shoots them.

6.2.3. Exercise: A Monte Carlo Shooting Scheme

From Apollodorus:

Hercules scared the birds. They could not abide the sound, but fluttered up in a fright, and in that way Hercules shot them.

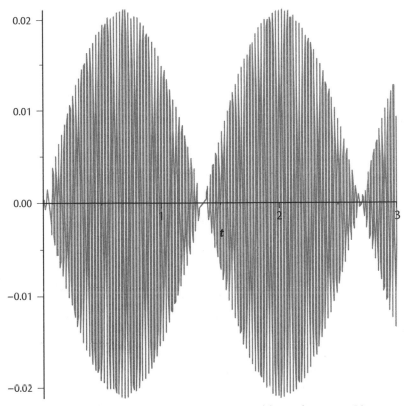

Figure 6.4. The solution to the castanet problem when $\omega = 80\pi$.

TASK: We expect that after scaring the birds into taking flight, Hercules will shoot them down. Imagine that he has forced them to fly in a certain direction and that their flight path takes them in front of a large square marked in the trees by a rope. Inscribed inside the square is a circle. Hercules stands perpendicular to the birds' flight path and shoots his arrows at them as they cross in front of him (see Figure 6.1). Although this task is not explicitly described by Apollodorus, perhaps Hercules is trying to impress the gods by approximating π.

SOLUTION: There is no doubt that Hercules hits his target. He's Hercules! As each bird flies in front of the square target, Hercules unleashes an arrow which hits its mark. According to Apollodorus, Hercules was taught to shoot with the bow by Eurytus. There are several instances in the myths about the legendary archery skills of our hero. Eurytus was the king of Oechalia, a town in Thessaly, and his grandfather was Apollo, the archer-god. Eurytus taught Hercules well the art of archery. According to Homer, Eurytus became so proud of his archery skills that he challenged Apollo. The god killed Eurytus because of his presumption, and Eurytus' bow was passed on to his son Iphitus, who later gave it to his friend Odysseus. It was this bow that Odysseus used to killed the suitors who wanted to take his wife, Penelope. But that is another famous myth.

Hercules does not need to complete this exercise in order to perform the Sixth Labor. The key concept here is that although all of the birds fly in front of the square, into Hercules' target area, their actual position inside the area is random. Some fly in front of the circle; others do not and stay at the edges of the square. This is indicative of a Monte Carlo experiment. In a typical Monte Carlo process, we start with a defined region C (in our case, a circle) and compute the number of randomly generated points in a set S (the square) that lie inside the inscribed circle C. The ratio of the number of points in C to the total number of points in S, typically denoted as ρ, is equal to the ratio of the two areas.

How does Hercules approximate π? The area of a circle is πr^2, where r is the radius of the circle. The circle is inside a square whose area is thus $(2r)^2 = 4r^2$. The ratio of these two areas is then

$$\rho = \frac{\pi r^2}{4r^2} = \frac{\pi}{4},$$

or $\pi = 4\rho$. So, as the Stymphalian birds fly in front of Hercules, he shoots them all, keeping track of those shot inside the circle. If Hercules tried this on a different day, the ratio ρ would be slightly different, even if the number of birds was the same. Why? Because the birds might take different flight paths. This is a process involving the generation of random numbers. Also, we would expect that a higher number of data points (in our case, more birds) would lead to greater accuracy in the approximation. Once Hercules empirically computes ρ by comparing the number of birds shot while inside the circle to the total number shot inside the square, he simply multiplies ρ by 4 to get π. Great Zeus!

By the way, the first Greek to try to develop a relationship between a circle and a square was Anaxagoras of Clazomenae, who lived

```
Microsoft Excel - Monte Carlo Pi.xls
 File   Edit   View   Insert   Format   Tools   Data   Window   Help   Adobe PDF
 ...                                                        Σ  ...  100%  ...  Aria
 ... Reply with Changes... End Review...

 J5        fx  =4*J2/(J2+J3)
```

	A	B	C	D	E	F	G	H	I	J	
1		Trial	x	y	x^2 + y^2	Inside?					
2		1	0.43410	0.07818	0.19456	Yes			Yes:	7783	
3		2	0.18880	0.26557	0.10617	Yes			No:	2217	
4		3	0.47437	0.14962	0.24741	Yes					
5		4	0.80405	0.40832	0.81323	Yes			Pi:	3.1132	
6		5	0.44339	0.28558	0.27815	Yes					
7		6	0.22752	0.70112	0.54334	Yes					
8		7	0.94521	0.25250	0.95718	Yes					
9		8	0.17720	0.77789	0.63651	Yes					
10		9	0.00792	0.21780	0.04750	Yes					
11		10	0.82775	0.36149	0.81586	Yes					
12		11	0.26489	0.92326	0.92257	Yes					
13		12	0.52611	0.90016	1.24340	No					
14		13	0.47222	0.52236	0.49585	Yes					
15		14	0.16380	0.87999	0.80122	Yes					
16		15	0.49302	0.27917	0.32101	Yes					
17		16	0.07070	0.71300	0.00732	Yes					
18		17	0.35648	0.70665	0.62643	Yes					
19		18	0.27491	0.31237	0.17315	Yes					
20		19	0.18464	0.96219	0.95990	Yes					
21		20	0.02932	0.36356	0.13304	Yes					
22		21	0.36793	0.46428	0.35093	Yes					
23		22	0.11561	0.51750	0.28118	Yes					
24		23	0.38885	0.50171	0.40292	Yes					
25		24	0.09288	0.73459	0.54825	Yes					
26		25	0.28653	0.97306	1.02894	No					
27		26	0.71449	0.29779	0.59918	Yes					
28		27	0.25968	0.67939	0.52900	Yes					
29		28	0.52735	0.90529	1.09766	No					
30		29	0.67250	0.16079	0.47811	Yes					
31		30	0.19449	0.80511	0.68603	Yes					
32		31	0.08599	0.05018	0.00991	Yes					
33		32	0.00632	0.29607	0.08770	Yes					
34		33	0.05791	0.22182	0.05256	Yes					
35		34	0.60803	0.31782	0.47071	Yes					
36		35	0.92742	0.84233	1.56962	No					
37		36	0.78305	0.70772	1.11403	No					
38		37	0.99518	0.66893	1.43786	No					
39		38	0.08595	0.70236	0.50069	Yes					
40		39	0.98917	0.26060	1.04637	No					
41		40	0.43065	0.58315	0.52552	Yes					
42		41	0.69345	0.00840	0.48094	Yes					
43		42	0.50115	0.58793	0.59681	Yes					
44		43	0.14002	0.81513	0.68404	Yes					
45		44	0.58062	0.81788	1.00606	No					
46		45	0.02936	0.05629	0.00403	Yes					
47		46	0.35079	0.34473	0.24189	Yes					

Figure 6.5. A spreadsheet method to approximate π.

about 60 years after depictions of the legends of Hercules first started appearing on vases and in other Greek art. While in prison, Anaxagoras supposedly developed a method for drawing a square which had the same area as a circle.

You can try this. Take a large square and draw a circle inside it. Make sure the circle's radius is exactly half the length of one side of the square. Then take a suction dart gun, stand a few feet away from your target, aim at the square, and shoot. Hopefully, all of the darts will land inside the square. Many of the darts should land inside the inscribed circle. You have to be far enough away to generate some randomness in the experiment. In this way, you can approximate π.

If you're not a great artist, or you don't want to shoot suction darts at your wall, type "Monte Carlo Pi Applets" into your computer's internet search engine and conduct an online experiment. The applets you find will allow you to "throw" tens of thousands of darts, each of whose position is generated randomly, into the square. Start with 100 or 200 and keep track of the approximation as the total number as you continue to add data points. As the number of darts increases, the approximation should get closer to $\pi \approx 3.1415\ldots$.

As you might imagine, even tens of thousands of randomly generated data points will not give a completely accurate representation of π; remember, this is an approximation. Also, by the Law of Large Numbers, the error from π will be on the order of $n^{-1/2}$ as the number of points n gets large. Monte Carlo simulations are very slow to converge. You can also write your own scheme using a spreadsheet or calculator. Generate (x, y) pairs of random numbers, each between -1 and 1. See Figure 6.5 for an example.

Count the number of (x, y) pairs whose sum of squares is less than or equal to 1 and divide by the total number of pairs. Multiply this ratio by 4, and you have just approximated π. It is easy to generate thousands of random (x, y) pairs. The ones whose sum of squares satisfies $x^2 + y^2 \leq 1$ will be inside the unit circle. The ratio is the same as before. Have patience. It works. Easy as π.

CHAPTER

7

The Seventh Labor: The Cretan Bull

From Apollodorus:

The seventh labour he enjoined on him was to bring the Cretan bull. Acusilaus says that this was the bull that ferried across Europa for Zeus; but some say it was the bull that Poseidon sent up from the sea when Minos promised to sacrifice to Poseidon what should appear out of the sea. And they say that when he saw the beauty of the bull he sent it away to the herds and sacrificed another to Poseidon; at which the god was angry and made the bull savage. To attack this bull Hercules came to Crete, and when, in reply to his request for aid, Minos told him to fight and catch the bull for himself, he caught it and brought it to Eurystheus, and having shown it to him he let it afterwards go free. But the bull roamed to Sparta and all Arcadia, and traversing the Isthmus arrived at Marathon in Attica and harried the inhabitants.

7.1. The Tasks

The first six labors were confined to the Peleponnesus. Beginning with the Seventh Labor, Hercules finds he must travel away from his home to Crete. He has two tasks to complete. The first task involves the actual riding of the Cretan bull. Hercules must stay on the savage beast for at least 8 seconds. Using actual data from the official web site of Professional Bull Riders, Inc., Hercules must tackle some probability problems in the **Riding the Bull** problem. The second task also deals with probability. Once the bull is released, it attacks the residents of Attica according to a specified probability density function. Help Hercules calculate the chance of an attack in **The Marathon Attacks** problem.

7.1.1. Exercise: Riding the Bull

TASK: The first task needed to accomplish this labor involves riding the bull. Suppose that 1886 men from Crete randomly volunteer to try to ride the Cretan bull, but only 268 can stay on the bull for at least 8 seconds. The rest are bucked off of the bull. Let p represent the proportion of all riders who stay on the bull for at least 8 seconds. Estimate p (specifically, find a point estimate) and determine a 95 percent confidence interval for p. What is the meaning of this confidence interval?

7.1.2. The Marathon Attacks

TASK: Hercules is successful in taking the Cretan bull to Eurystheus. However, once the bull is set free, it attacks inhabitants in and around Attica with a fury. The further the bull is from Marathon, the fiercer its demeanor (and the less safe the people), and the likelier it becomes that the bull will attack someone. Let's suppose the probability density function that models the chance that the bull will attack someone within t kilometers of Marathon is given by $a(t) = \frac{9}{4}te^{-3t/2}$. Answer the following two questions:

1. What is the probability that the bull will attack someone within 2 kilometers of Marathon?
2. Within what distance of Marathon will the probability that the bull will attack someone be 90 percent?

7.2. The Solutions

7.2.1. Exercise: Riding the Bull

From Apollodorus:

> To attack this bull Hercules came to Crete, and when, in reply to his request for aid, Minos told him to fight and catch the bull for himself, he caught it and brought it to Eurystheus.

TASK: The first task needed to accomplish this labor involves riding the bull. Suppose that 1886 men from Crete randomly volunteer to try to ride the Cretan bull, but only 268 can stay on the bull for at least 8 seconds. The rest are bucked off of the bull. Let p represent the proportion of all riders who stay on the bull for at least 8 seconds. Estimate p (specifically, find a point estimate) and determine a 95 percent confidence interval for p. What is the meaning of this confidence interval?

SOLUTION: We will use modern data, which hopefully provides a suitable estimate of how riders performed in the past. The following passage is taken from www.pbrnow.com, the official web site of Professional Bull Riders, Inc., (PBR):

> The total score possible for a bull ride is 100 points. Half of that total is based on the performance of the bull and how difficult he is to ride. Judges look for bulls with speed, power, drop in the front end, kick in the back end, directions changed and body rolls. A body roll occurs when a bull is in the air and kicks either his hind feet or all four feet to the side. The more of these characteristics a bull displays during a ride, the higher the mark is for the bull. Judges are allowed to award a cowboy a re-ride if they feel the bull did not perform at the level of other bulls in the competition and, therefore did not give the rider a fair chance to earn a high score.
>
> The other half of the ride is determined by the rider's ability to match the moves of the bull beneath him. Judges look for constant control and good body position throughout the ride. Spurring the bull is not required but extra "style points" are awarded for doing so. The rider must stay aboard the bull for eight seconds. The clock begins when the bull's shoulder or hip crosses the plane of the bucking chutes and stops when the bull rider's hand comes out of the rope or he touches the ground. The bull rider must ride with one hand and is disqualified if he touches himself or the bull during the eight-second ride.

Bulls are scored every ride, regardless of whether the rider makes the 8 seconds. Each of the four judges can award a maximum of 25 points for the bull and 25 points for the rider. The four judges' scores are added and then divided by 2. The top 100 bulls currently being used by the

PBR have an astonishing 85.79 percent career "buckoff" percentage, meaning that out of a total of 1886 career attempts (as of July 2007), these bulls have been ridden for the full 8 seconds only 268 times! It's not easy to ride a bull. According to the 2007 Professional Bull Riders statistics page, the top 19 riders on tour have demonstrated a 48.49 percent chance of staying on the bull for the whole 8 seconds. Riders ranked lower than 19 have a more difficult time staying on the bull (note the above 268 out of 1886). When the top riders do get credit for a ride, their average score is 85.64 out of a possible 100.

So, what about our hero and the Cretan bull? Either a rider stays on the bull for 8 seconds (success) or he does not (failure). This is known as a *binomial trial*. The proportion of successful riders in the population is found by dividing the number of successes by the number of trials. We call this a point estimate, designated by $\hat{p}$, so $\hat{p} = 268/1886 \approx 0.1421$. Just over 14 percent of the riders in the population are successful in riding the bull for at least 8 seconds.

The 95 percent confidence interval for p is given by

$$\hat{p} - E < p < \hat{p} + E,$$

where E is the margin of error determined by the formula

$$E \approx z_c \sqrt{\frac{\hat{p}(1 - \hat{p})}{n}},$$

where c is the confidence level (between 0 and 1 and in this case, 0.95), n is the number of samples, and z_c is the critical value for the confidence level c determined from the standard normal distribution. We usually find z_c in a table of confidence interval critical values (any standard statistics book has such a table). In particular, $z_{0.95} = 1.96$, so

$$E \approx 1.96 \sqrt{\frac{(0.1421)(0.8579)}{1886}} \approx 0.0158.$$

The confidence interval is therefore

$$\hat{p} - E < p < \hat{p} + E,$$
$$0.1421 - 0.0158 < p < 0.1421 + 0.0158,$$
$$0.1263 < p < 0.1579.$$

What does this interval mean? If we computed an interval for each of the 1886 riders, we would find that 95 percent of the intervals contained p, the proportion of all riders who stay on the bull for at least 8 seconds. In other words, we can be 95 percent sure that between 12.63 percent and 15.79 percent of the riders will stay on the bull for at least 8 seconds. This makes sense.

As the confidence interval increases, so does the margin of error. Of course, Hercules is able to stay on the Cretan bull, but it takes much more than 8 seconds to tame it, as we will see next.

7.2.2. The Marathon Attacks

From Apollodorus:

> But the bull roamed to Sparta and all Arcadia, and traversing the Isthmus arrived at Marathon in Attica and harried the inhabitants.

TASK: Hercules is successful in taking the Cretan bull to Eurystheus. However, once the bull is set free, it attacks inhabitants in and around Attica with a fury. The further the bull is from Marathon, the fiercer its demeanor (and the less safe the people), and the likelier it becomes that the bull will attack someone. Let's suppose the probability density function that models the chance that the bull will attack someone within t kilometers of Marathon is given by $a(t) = \frac{9}{4}te^{-3t/2}$. Answer the following two questions:

1. What is the probability that the bull will attack someone within 2 kilometers of Marathon?
2. Within what distance of Marathon will the probability that the bull will attack someone be 90 percent?

SOLUTION: (1) We are given that the probability density function that models the chance of an attack on the local inhabitants based on the distance from Marathon is

$$a(t) = \frac{9}{4}te^{-3t/2}.$$

First, since the domain for the probability density function is $t \geq 0$, we leave it to the reader to show that this is indeed a probability density function (integrate $a(t)$ from 0 to infinity—it must integrate to 1, which is the sum of all the probabilities). In order to calculate the probability that an attack will take place within 2 kilometers of Marathon, we must integrate the probability density function from 0 to 2 kilometers. In doing so, we find

$$\int_0^2 \left(\frac{9}{4} t e^{-3t/2} \right) dt = -\frac{3t+2}{2} e^{-3t/2} \Big|_0^2 = 1 - 4e^{-3} \approx 0.8008.$$

There is a 80.08 percent chance that the bull will attack someone within 2 kilometers of Marathon. As the bull gets closer to Marathon (t decreases), the probability of attack decreases (there is less area under the curve of $a(t)$).

(2) To answer the second question, we integrate the attack function and solve for the upper limit that makes the value of the integral equal to 0.90. We begin with

$$\int_0^b \left(\frac{9}{4} t e^{-3t/2} \right) dt = \left(-\frac{3t+2}{2} e^{-3t/2} \right) \Big|_0^b.$$

Evaluating the limits and setting the expression equal to 0.90, we find that

$$1 - \frac{3b+2}{2} e^{-3b/2} = 0.90.$$

So, this problem comes down to solving $\frac{1}{2}(3b+2)e^{-3b/2} = 0.10$. Now, we employ a root-finding technique, such as Newton's Method, to solve this equation for b. Let $f(t) = (3t+2)e^{-3t/2} - 0.20$, so that our task is to solve $f(t) = 0$. We start with an initial guess, say $t_0 = 2$, since we know that within 2 kilometers, there is a high probability that the bull will attack. Newton's Method takes the following recursive form:

$$t_{n+1} = t_n - \frac{f(t_n)}{f'(t_n)}. \tag{7.1}$$

We know t_0, and we can evaluate $f(t_0)$, so we need to find $f'(t)$ and evaluate it at t_0. Differentiating, we find that $f'(t) = -\frac{9t}{2}e^{-3t/2}$. Then

$$
\begin{aligned}
t_1 &= t_0 - \frac{f(t_0)}{f'(t_0)} \\
&= 2 - \frac{(3 \times 2 + 2)e^{-3 \times 2/2} - 0.20}{-\frac{9 \times 2}{2}\,e^{-3 \times 2/2}} \\
&\approx 2.44.
\end{aligned}
$$

Now we use t_1 to find the next iteration, t_2.

$$
\begin{aligned}
t_2 &= 2.44 + \frac{(3 \times 2.44 + 2)\,e^{-3 \times 2.44/2} - 0.20}{-\frac{9 \times 2.44}{2}\,e^{-3 \times 2.44/2}} \\
&\approx 2.58.
\end{aligned}
$$

Continuing the recursive technique, we find that (to two decimal places) $t_3 \approx 2.59 \approx t_4 \approx t_5 = \cdots$. This means that the root to $f(t)$ converges to $t = 2.59$ and that within 2.59 kilometers of Marathon, the probability that the bull will attack the local population is 90 percent. You can find plenty of Newton's Method applets online to numerically approximate the solution (root) to an equation.

According to Diodorus, after completing the labor of the Cretan bull,

> Hercules established the Olympic Games, having selected for so great a festival the most beautiful of places, which was the plain lying along the banks of the Alpheus River, where he dedicated these Games to Zeus the Father.

However, that is another story. . . .

The Eighth Labor: The Horses of Diomedes

From Apollodorus:

The eighth labour he enjoined on him was to bring the mares of Diomedes the Thracian to Mycenae. Now this Diomedes was a son of Ares and Cyrene, and he was king of the Bistones, a very warlike Thracian people, and he owned man-eating mares. So Hercules sailed with a band of volunteers, and having overpowered the grooms who were in charge of the mangers, he drove the mares to the sea. When the Bistones in arms came to the rescue, he committed the mares to the guardianship of Abderus, who was a son of Hermes, a native of Opus in Locris, and a minion of Hercules; but the mares killed him by dragging him after them. But Hercules fought against the Bistones, slew Diomedes and compelled the rest to flee. And he founded a city Abdera beside the grave of Abderus who had been done to death, and bringing the mares he gave them to Eurystheus. But Eurystheus let them go, and they came to Mount Olympus, as it is called, and there they were destroyed by the wild beasts.

8.1. The Tasks

Thrace lies on the northern shores of the Aegean Sea, in what was the eastern part of Macedonia. Diomedes kept four savage mares, to which he fed the flesh of unsuspecting strangers. It is said that he was as savage as his mares; they were totally uncontrollable and were tethered by chains to a bronze manger. Hercules has three tasks to complete. After capturing the mares and taking them from their stables, he drives them to the sea. At the same time, the warlike Bistones make their way to the sea to intercept him. Hercules must determine how fast the distance between him and the Bistones is changing (the **Driving the Mares to the Sea** problem). Then Hercules engages in a battle with the Bistones and Diomedes. He uses a slingshot to disable Diomedes, and so he must describe the motion of the stone in the sling parametrically (the **Hercules' Slingshot** problem). Before he takes the mares to Mycenae, Hercules founds the city of Abdera. The third task involves modeling the population of Abdera (in **The City of Abdera** problem).

8.1.1. Driving the Mares to the Sea

TASK: Hercules overpowers the grooms who are guarding Diomedes' horses, frees the four horses from their chains, and drives them to a port by the sea which is 2 kilometers due south of the stables. One of the grooms sends out an alarm to where the Bistones are living, which is due east of the mares' stables. The Bistones begin racing on horseback westward toward the stables in order to stop Hercules and the mares. The speed of the Bistones is 7 meters per second at the moment they are 600 meters east of the stables. At the same moment, Hercules and the mares are traveling at 4.5 meters per second and are 400 meters south of the stables. Determine how fast the distance between Hercules and the Bistones is changing at that instant. In addition, if the Bistones first travel west all the way to the stables and then south to the port, will they overcome Hercules before he reaches the sea with the mares?

8.1.2. Hercules' Slingshot

TASK: While Hercules is fighting the Bistones, he disables Diomedes by using a slingshot. He swings a stone in his sling at a constant speed in a circle parallel to the ground and centered over his head, 2 meters off of the ground. Assume the stone travels in a counterclockwise direction (counterclockwise according to Zeus, who is looking downward at Hercules). At time $t = 0$, the stone is positioned at Hercules' right side, 1.5 meters from the center of his head. After the stone travels a full circle plus a quarter, it leaves the sling. Describe the circular trajectory of the stone's motion. Determine the velocity and acceleration components of the stone just as it leaves the sling. How would you determine the position of the stone after it is airborne?

8.1.3. Exercise: The City of Abdera

TASK: For 100 years after its founding the population growth of Abdera followed a logistics-type curve. Suppose that when the city had 100 inhabitants, its relative birth rate was 17 percent and its relative death rate was only 7 percent. By the time the population had tripled to 300, the relative birth rate and death rate were 14 percent and 9 percent, respectively. Assume that the relative growth rate g of the population of Abdera (which is the difference between the birth rate and mortality rate) is a linearly decreasing function of the current number of inhabitants y: $g = k - ry$, for some constants k and r. Determine and plot the long-range population growth of the city and find the equilibrium value if one exists. In addition, if the population of Abdera suddenly surges to 550 (perhaps because of refugees arriving), what effect will this have on the long-term population growth?

8.2. The Solutions

8.2.1. Driving the Mares to the Sea

From Apollodorus:

The eighth labour he enjoined on him was to bring the mares of Diomedes the Thracian to Mycenae. Now this Diomedes was a son of Ares and Cyrene, and he was king of the Bistones, a very warlike Thracian people, and he owned man-eating mares. So Hercules sailed with a band of volunteers, and having overpowered the grooms who were in charge of the mangers, he drove the mares to the sea.

TASK: Hercules overpowers the grooms who are guarding Diomedes' horses, frees the four horses from their chains, and drives them to a

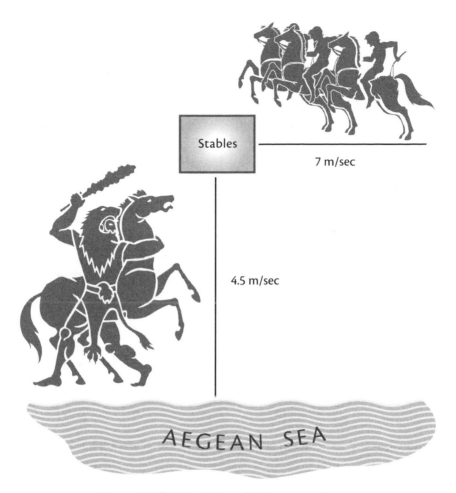

Figure 8.1. Driving Diomedes' horses to the sea.

port by the sea which is 2 kilometers due south of the stables. One of the grooms sends out an alarm to where the Bistones are living, which is due east of the mares' stables. The Bistones begin racing on horseback westward toward the stables in order to stop Hercules and the mares. The speed of the Bistones is 7 meters per second at the moment they are 600 meters east of the stables. At the same moment, Hercules and the mares are traveling at 4.5 meters per second and are 400 meters south of the stables (see Figure 8.1). Determine how fast the distance between Hercules and the Bistones is changing at that instant. In addition, if the Bistones first travel west all the way to the stables and then south to the port, will they overcome Hercules before he reaches the sea with the mares?

SOLUTION: Hercules wants to drive the horses from the stables to the sea as fast as possible. Let $x(t)$ be the distance at time t from the mares' stables to where the Bistones are situated. Let $y(t)$ be the distance at time t from the stables to the sea. Let $z(t)$ be the distance between Hercules and the Bistones at any time t; all distances are measured in meters. We are given that Hercules and the mares travel at a speed of 4.5 meters per second (this is dy/dt), while the Bistones travel at a speed of 7 meters per second (this is dx/dt). We will assume that the initial directions traveled by Hercules and by the groom remain at right angles, which allows us to use the Pythagorean Theorem to write

$$x(t)^2 + y(t)^2 = z(t)^2.$$

We differentiate each term with respect to t, obtaining

$$2x\frac{dx}{dt} + 2y\frac{dy}{dt} = 2z\frac{dz}{dt}.$$

At the moment in question, $x = 600$ meters and $y = 400$ meters. This means that $z = \sqrt{600^2 + 400^2} \approx 721.11$. Now we can substitute in what we know and solve for dz/dt:

$$(600)(-7) + (400)(4.5) = 721.11\frac{dz}{dt}.$$

Note that dx/dt is negative, as the distance from the Bistones to the stables is decreasing. Solving for dz/dt, we have

$$\frac{dz}{dt} = \frac{(600)(-7) + (400)(4.5)}{721.112} \approx -3.328 \text{ meters per second.}$$

The distance between Hercules and the groom is decreasing at a rate of about 3.3 meters per second, so the Bistones are gaining on Hercules.

Now, on to the question of whether the Bistones will catch Hercules and the mares. The Bistones must travel a total of $600 + 2000 = 2600$ meters, while Hercules travels only $2000 - 400 = 1600$ meters. At their rate of 7 meters per second, the Bistones need 371.4 seconds, or about 6 minutes and 11 seconds, to reach the port. At the rate of 4.5 meters per second, Hercules needs 355.6 seconds, or about 5 minutes and 56 seconds, to reach the port. Hercules arrives with less than $\frac{1}{2}$ minute to spare, giving him time to turn the horses over to Abderus and fight the Bistones. Lacking in experience, the young Abderus is unfortunately eaten by the savage beasts.

8.2.2. Hercules' Slingshot

From Apollodorus:

> The eighth labour he enjoined on him was to bring the mares of Diomedes the Thracian to Mycenae.... Hercules fought against the Bistones, slew Diomedes and compelled the rest to flee.

TASK: While Hercules is fighting the Bistones, he disables Diomedes by using a slingshot. He swings a stone in his sling at a constant speed in a circle parallel to the ground and centered over his head, 2 meters off of the ground. Assume the stone travels in a counterclockwise direction (counterclockwise according to Zeus, who is looking downward at Hercules). At time $t = 0$, the stone is positioned at Hercules' right side, 1.5 meters from the center of his head. After the stone travels a full circle plus a quarter, it leaves the sling. Describe the circular trajectory of the stone's motion. Determine the velocity and acceleration components of the stone just as it leaves the sling. How would you determine the position of the stone after it is airborne?

SOLUTION: We consider the point where Hercules is standing as the origin, with his right and front sides being the positive x- and y-axes, respectively. A stone 2 meters off the ground directly over his head has coordinates $(x, y, z) = (0, 0, 2)$. At $t = 0$, the stone is at $(x, y, z) = (1.5, 0, 2)$. Assuming the sling is fully extended as it and the stone circle Hercules' head, and further assuming that the sling rotates in a horizontal plane at a constant value of z, we can say that the radius of the sling is $r = 1.5$ meters. So, with $x = 1.5$, $y = 0$, and $z = 2$, we can parametrize the sling's motion as

$$x(t) = 1.5 \cos(t),$$
$$y(t) = 1.5 \sin(t), \qquad\qquad (8.1)$$
$$z(t) = 2.$$

The stone will return to the same position at $t = 2n\pi$ for every integer n. After Hercules swings the sling a full circle plus a quarter, the position of the stone is $(x, y, z) = (0, 1.5, 2)$. At that point, $t = \frac{5\pi}{2}$. From Equation 8.2, the position vector is

$$\mathbf{x}(t) = 1.5 \ \cos(t)\mathbf{i} + 1.5 \sin(t)\mathbf{j} + 2\mathbf{k}.$$

Differentiating, we find the velocity vector $\mathbf{v}(t)$ as the stone leaves the sling to be

$$\mathbf{v}(t) = -1.5 \ \sin(t)\mathbf{i} + 1.5 \ \cos(t)\mathbf{j} + 0\mathbf{k},$$

and at $t = \frac{5\pi}{2}$,

$$\mathbf{v}\left(\frac{5\pi}{2}\right) = -1.5 \ \sin\left(\frac{5\pi}{2}\right)\mathbf{i} + 1.5 \cos\left(\frac{5\pi}{2}\right)\mathbf{j} = -1.5\mathbf{i}.$$

The acceleration vector $\mathbf{a}(t)$ is found by differentiating $\mathbf{v}(t)$ at $t = \frac{5\pi}{2}$:

$$\mathbf{a}\left(\frac{5\pi}{2}\right) = -1.5 \cos\left(\frac{5\pi}{2}\right)\mathbf{i} - 1.5 \sin\left(\frac{5\pi}{2}\right)\mathbf{j} = -1.5\mathbf{j}.$$

We can now describe the movement of the airborne stone using a system of differential equations. Since the stone is affected by the force of gravity only after it leaves the sling, we have

$$\frac{d^2x}{dt^2} = 0, \qquad \frac{d^2y}{dt^2} = 0, \qquad \frac{d^2z}{dt^2} = -g. \qquad (8.2)$$

This is a system of second-order differential equations requiring two conditions for each variable. They are

$$x(0) = 1.5, \qquad y(0) = 0, \qquad z(0) = 2$$

and

$$\frac{dx}{dt}(0) = -1.5, \qquad \frac{dy}{dt}(0) = 0, \qquad \frac{dz}{dt}(0) = 0.$$

Determining the solution to the system of equations is left as an exercise for the reader.

The stone flies out of the sling and strikes Diomedes, knocking him unconscious. Then Hercules finds young Abderus torn to pieces by the four horses, and knowing the brutality King Diomedes has caused, Hercules takes him to the great mares and throws him into their pen (according to legend), whereupon the four mares proceed to devour their own master. This causes the horses to become calm and subdued, which makes it easy for Hercules to bind their mouths shut and drive them back to King Eurystheus. Hercules founded the city of Abdera near Abderus' tomb, where athletic games consisting of boxing, pancratium, and wrestling were held in honor of young Abderus. As a note of interest, Bucephalus, Alexander the Great's horse and arguably the most famous horse of antiquity, is said to be descended from these man-eating mares.

8.2.3. Exercise: The City of Abdera

From Apollodorus:

> And he founded a city Abdera beside the grave of Abderus who had been done to death....

TASK: For 100 years after its founding, the population growth of Abdera followed a logistics-type curve. Suppose that when the city had 100 inhabitants, its relative birth rate was 17 percent and its relative death rate was only 7 percent. By the time the population had tripled to 300, the relative birth and death rates were 14 percent and 9 percent, respectively. Assume that the relative growth rate g of the population of Abdera (which is the difference between the birth and mortality rates) is a linearly decreasing function of the current number of inhabitants y: $g = k - ry$, for some constants k and r. Determine and plot the long-range population growth of the city and find the equilibrium value if one exists. In addition, if the population of Abdera suddenly surges to 550 (perhaps because of refugees arriving), what effect will this have on the long-term population growth?

SOLUTION: Abdera was an actual town on the coast of Thrace near the mouth of the Nestos River. After Hercules supposedly founded it, its population continued to grow, and there are records showing that it was actually quite prosperous in 544 BC, when the majority of the people of Teos migrated to Abdera to escape the Persians. The town seems to have declined in importance after the middle of the fourth century. Many sources cite a notion that the air of Abdera was said to cause stupidity. The ruins of the town may still be seen today; they cover seven small hills and extend from an eastern to a western harbor.

Define $y(t)$ as the population of Abdera in year t. The relative growth rate is the ratio of the change in population to the population itself. Symbolically, this is

$$g = \frac{\frac{dy}{dt}}{y} = \frac{1}{y}\frac{dy}{dt}.$$

We are given that this relative growth rate is a linearly decreasing function of the current number of inhabitants. Therefore,

$$g = \frac{1}{y}\frac{dy}{dt} = k - ry,$$

where k and r are constants. Note that the right-hand side represents a decreasing line. The population is in equilibrium when there is no growth (either positive or negative), so $g = 0$ or

$$\frac{dy}{dt} = y(k - ry) = 0.$$

Solving for y, we find equilibrium values when $y = 0$ or when $y = k/r$. Let's define the carrying capacity M of the population to be this limiting value of y, so $M = k/r$. Physically, M is the maximum population that the environment can sustain. Putting all this together, the logistics equation for the population $y(t)$ appears in the form

$$\frac{dy}{dt} = ky \left(1 - \frac{y}{M} \right). \tag{8.3}$$

Here k is the growth constant and M is the carrying capacity of the population.

We need to determine what the growth constant and carrying capacity are. We are given that when $y = 100$, the relative birth rate was 17 percent and the relative death rate was only 7 percent. This means that the growth rate is equal to $g_{100} = 0.17 - 0.07 = 0.10$, or 10 percent. Further, when $y = 300$, we can determine that $g_{300} = 0.14 - 0.09 = 0.05$, or 5 percent. The city shows positive growth of the population, although this growth does slow down as the population gets larger. Note that at this point, we don't know *when* the populations were at 100 or 300, but we can still obtain a general solution for the differential equation. We substitute these growth rates with $g = k - ry$ to obtain two equations with two unknowns:

$$0.10 = k - 100r \qquad \text{and} \qquad 0.05 = k - 300r,$$

with solutions $k = 1/8$ and $r = 1/4000$. This leads to $M = 4000/8 = 500$. The city of Abdera can sustain a population of 500 inhabitants. This is a stable equilibrium value. As long as the population remains below 500 inhabitants, it will grow to 500.

We can find an explicit formula for $y(t)$ by solving

$$\frac{dy}{y(M - y)} = \frac{k}{M} dt \tag{8.4}$$

using separation of variables. We can rewrite the left-hand side as

$$\frac{1}{y(M - y)} = \frac{1}{M} \left(\frac{1}{y} + \frac{1}{M - y} \right).$$

Integrating Equation 8.4,

$$\int \frac{1}{M} \left(\frac{1}{y} + \frac{1}{M-y} \right) dy = \int \frac{k}{M} \, dt.$$

Multiplying by M and integrating yields

$$\ln|y| - \ln|M-y| = kt + C$$

for some constant C. We can multiply this equation through by (-1) to simplify the left-hand side and get

$$\ln \left| \frac{M-y}{y} \right| = -kt - C.$$

Exponentiating both sides yields

$$\left| \frac{M-y}{y} \right| = e^{-kt-C} = e^{-C} e^{-kt} = c \, e^{-kt},$$

where $c = e^{-C}$. We can let c be positive or negative, which allows us to drop the absolute value signs and solve for y. Now we have

$$\frac{M-y}{y} = \frac{M}{y} - 1 = ce^{-kt},$$

so

$$y = \frac{M}{1 + ce^{-kt}}.$$

Substituting in the values of k and M, we found above that

$$y = \frac{500}{1 + ce^{-t/8}}. \qquad (8.5)$$

Equation 8.5 gives the general solution. Knowing an initial condition will allow us to solve for c. For example, if, after 10 years, Abdera is found to have 50 inhabitants, then a particular solution satisfying the

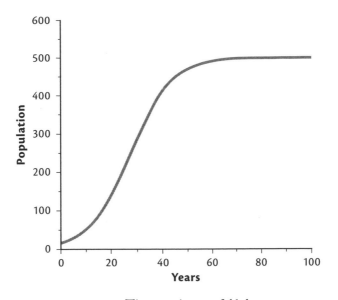

Figure 8.2. The population of Abdera.

condition of $y(10) = 50$ would be

$$y = \frac{500e^{-5/4}}{e^{-5/4} + 9e^{-t/8}}.$$

However, we can see a general trend of population growth in Figure 8.2. The population increases with time up to the carrying capacity of 500, which is the stable (or attracting) equilibrium value. If there are suddenly 550 inhabitants in Abdera and the growth rate does not change, we would expect the population of the city to decrease to the equilibrium value of 500. The environment cannot sustain more than 500 inhabitants.

8.3. The Diomedes Sudoku Puzzle

After he had completed this labor, Hercules sailed forth with Jason and the Argonauts as a member of the expedition to find the golden fleece. But that is another famous myth.... Before we move on, however, you have completed eight labors. Time for celebration! Enjoy the following

		3	2					
		2				9	4	7
		6						1
				9				
7						2		
6	1	4				5		
					9	1		

Figure 8.3. Diomedes sudoku.

sudoku puzzle in Figure 8.3. This is a "knight" sudoku puzzle. The rules are the same as in normal sudoku, except that there is an extra condition that cells which are a knight's chess move apart (either two left/right and one up/down or two up/down and one left/right) must be different. Other than that restriction, each box must contain the numbers 1 through 9 only once, and each column or row may have only the numbers 1 through 9 once. Good luck!

The Ninth Labor: The Belt of Hippolyte

From Apollodorus:

The ninth labour he enjoined on Hercules was to bring the belt of Hippolyte. She was queen of the Amazons, who dwelt about the river Thermodon, a people great in war; for they cultivated the manly virtues, and if ever they gave birth to children through intercourse with the other sex, they reared the females; and they pinched off the right breasts that they might not be trammelled by them in throwing the javelin, but they kept the left breasts, that they might suckle. Now Hippolyte had the belt of Ares in token of her superiority to all the rest. Hercules was sent to fetch this belt because Admete, daughter of Eurystheus, desired to get it. So taking with him a band of volunteer comrades in a single ship he set sail and put in to the island of Paros, which was inhabited by the sons of Minos, to wit, Eurymedon, Chryses, Nephalion, and Philolaus. But it chanced that two of those in the ship landed and were killed by the sons of Minos. Indignant at this, Hercules killed the sons of Minos on the spot and besieged the rest closely, till they sent envoys to request that in the room of the murdered men he would take two, whom he pleased. So he raised the siege, and taking on board the sons of Androgeus, son of Minos, to wit, Alcaeus and Sthenelus, he came to Mysia, to the court of Lycus, son of Dascylus, and was entertained by him; and in a battle between him and the king of the Bebryces Hercules sided with Lycus and slew many, amongst others King Mygdon, brother of Amycus. And he took much land from the Bebryces and gave it to Lycus, who called it all Heraclea.

Having put in at the harbor of Themiscyra, he received a visit from
Hippolyte, who inquired why he was come, and promised to give him the
belt. But Hera in the likeness of an Amazon went up and down the multitude
saying that the strangers who had arrived were carrying off the queen. So
the Amazons in arms charged on horseback down on the ship. But when
Hercules saw them in arms, he suspected treachery, and killing Hippolyte
stripped her of her belt. And after fighting the rest he sailed away and
touched at Troy.

But it chanced that the city was then in distress consequently on the wrath
of Apollo and Poseidon. For desiring to put the wantonness of Laomedon to
the proof, Apollo and Poseidon assumed the likeness of men and undertook
to fortify Pergamum for wages. But when they had fortified it, he would not
pay them their wages. Therefore Apollo sent a pestilence, and Poseidon a
sea monster, which, carried up by a flood, snatched away the people of the
plain. But as oracles foretold deliverance from these calamities if Laomedon
would expose his daughter Hesione to be devoured by the sea monster, he
exposed her by fastening her to the rocks near the sea. Seeing her exposed,
Hercules promised to save her on condition of receiving from Laomedon the
mares which Zeus had given in compensation for the rape of Ganymede. On
Laomedon's saying that he would give them, Hercules killed the monster
and saved Hesione. But when Laomedon would not give the stipulated
reward, Hercules put to sea after threatening to make war on Troy.

And he touched at Aenus, where he was entertained by Poltys. And as
he was sailing away he shot and killed on the Aenian beach a lewd fellow,
Sarpedon, son of Poseidon and brother of Poltys. And having come to
Thasos and subjugated the Thracians who dwelt in the island, he gave it
to the sons of Androgeus to dwell in. From Thasos he proceeded to Torone,
and there, being challenged to wrestle by Polygonus and Telegonus, sons of
Proteus, son of Poseidon, he killed them in the wrestling match. And having
brought the belt to Mycenae he gave it to Eurystheus.

9.1. The Tasks

Several Greek heroes battled the Amazons, including Theseus,
Dionysus, and Achilles. Supposedly, the famous belt (called a girdle
in some accounts) was worn above the waist. If the wearer loosened it,
she would offer herself to a man; if the belt was taken forcibly, it was

considered rape. Hercules has three tasks to complete. First, he captures the sons of Minos and their men. He places his captives in a circle and then proceeds to kill every third man, continuing this process until only two men are alive. Help the prisoners determine where to stand if they hope to survive in **The Sons of Minos versus Hercules** problem. After this ordeal, Hercules attempts to get the Belt of Hippolyte. To thwart his efforts, however, the goddess Hera disguises herself as an Amazon warrior and travels "up and down the multitude" of the other Amazons, spreading the false rumor that Hercules and his band of strangers are carrying off the Amazon queen. Hercules must deal with the false rumor in **The Amazons and the Spread of a Rumor** problem. This task is followed by the **Hercules and the Kraken** problem, in which Hercules must determine the optimal distance at which the Greek crowd should view the sacrifice of Hesione to the mighty Kraken, a sea monster of giant proportions.

9.1.1. The Sons of Minos versus Hercules

TASK: After they are captured, the sons of Minos plead for their lives and the lives of their men. They convince Hercules to allow them to stand in a circle. Hercules will then kill every third man until only 2 are left, and those 2 will be allowed to go free. If Hercules has taken 41 men captive (including the sons of Minos), where should the 2 who hope to survive position themselves in the sequence?

9.1.2. The Amazons and the Spread of a Rumor

TASK: Hera, the wife of Zeus who wants Hercules to fail, disguises herself as an Amazon warrior. In this likeness, she travels "up and down the multitude" of the other Amazons, spreading the false rumor that Hercules and his band of strangers are carrying off the Amazon queen. Within 5 minutes, each Amazon warrior who hears the story passes it on to two other warriors who have not yet heard the rumor. Hera continues to spread the rumor as she passes down the ranks of the Amazons, telling a group of four Amazon women every 5 minutes. Assume she tells only four different groups of warriors and then disappears. How long will it take before 1500 warriors have heard the lie?

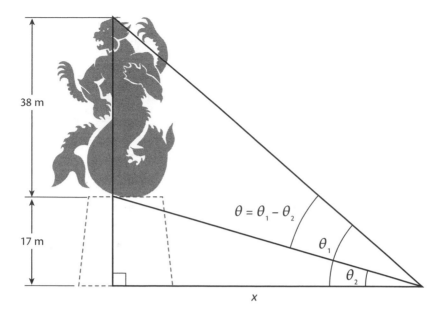

Figure 9.1. The view of the Kraken.

9.1.3. Exercise: Hercules and the Kraken

TASK: The Kraken was a terrifying sea monster who instilled fear in the ancient Greeks because of its size and appearance. It climbs onto the rocks at the edge of the shore and raises itself up before turning its attention to the princess Hesione, who is tied to the rocks. In order to instill the most fear, the Kraken ensures that its height is at a position where the angle of sight from the people is at a maximum. If the Kraken stands 38 meters high and climbs onto the rocks, which rise 17 meters above sea level, how far away do the villagers need to be to get the best (most fearful) view of the monster (see Figure 9.1)?

9.2. The Solutions

9.2.1. The Sons of Minos versus Hercules

From Apollodorus:

So taking with him a band of volunteer comrades in a single ship he set sail and put in to the island of Paros, which was inhabited by the sons of Minos, to wit, Eurymedon, Chryses, Nephalion, and Philolaus. But it chanced that two of those in the ship landed and were killed by the sons of Minos. Indignant at this, Hercules killed the sons of Minos on the spot and besieged the rest closely, till they sent envoys to request that in the room of the murdered men he would take two, whom he pleased.

TASK: After they are captured, the sons of Minos plead for their lives and the lives of their men. They convince Hercules to allow them to stand in a circle. Hercules will then kill every third man until only 2 are left, and those 2 will be allowed to go free. If Hercules has taken 41 men captive (including the sons of Minos), where should the 2 who hope to survive position themselves in the sequence?

SOLUTION: This is known as the *Josephus problem*, named after Flavius Josephus, a Jewish general and historian from the first century who recorded the destruction of Jerusalem in the year 70. After the Romans had captured the city of Jotapat, Josephus and 40 other Jews took refuge in a cave. The men resolved to kill themselves to avoid capture by the Romans. Josephus convinced his group to conduct mass suicide in an orderly manner, supposedly suggesting that they arrange themselves in a circle. Then every third man would be killed until the final one was left, who would then commit suicide. As the incident is described in his book, *The Jewish War*, Josephus delivered the following speech to the trapped men:

> Since we are resolved to die, come, let us leave the lot to decide the order in which we are to kill ourselves; let him who draws the first lot fall by the hand of him who comes next; fortune will thus take her course through the whole number, and we shall be spared from taking our lives with our own hands....

Each of the men selected then

> ...presented his throat to his neighbor, in the assurance that his general was forthwith to share his fate; for sweeter to them than life was the thought of death with Josephus.

Allegedly, Josephus and another man put themselves in two key positions, and they were the last 2 left alive. They then chose not to

commit suicide and were taken captive by the Romans. We hope to
show that after Hercules begins killing every third man in a group of
41, the last 2 left will occupy those same two key positions.

Before we tackle the 41-man problem, though, suppose that there are
only 6 men and that every third man is removed. Take a piece of paper
and write the numbers 1 through 6 in a circle. Start counting with 1 and
cross out every third number. You can easily verify that the sequence of
removed numbers is 3, then 6, then 4, then 2, then 5, leaving 1 as the
survivor (try other small cases for yourself).

Let n be the number of men in the circle. If we enumerate the
survivor position for the instances when $n = 6, 7, 8,$ or 9, we get

n	Positions in the circle								
6	3	6	4	2	5	1			
7	3	6	2	7	5	1	4		
8	3	6	1	5	2	8	4	7	
9	3	6	9	4	8	5	2	7	1

The last two positions in each row give the second-to-last and last
survivors. For our purposes, let's confine our algorithm to finding the
last survivor. Can we develop an algorithm to predict the survivor
position given an arbitrary number of men in the circle? Taking a large
circle of 41 positions and crossing out every third number takes time
and does not provide us with a general solution.

This is a nice place to discuss modulo (mod) arithmetic. Think of
a 24-hour clock (one day goes from midnight to midnight). When it
is 16:00, we know that it is really 4:00 PM. Ten hours later it is 2:00
AM. Normal addition has us add $16 + 10 = 26$, but time wraps around
at the end of the day, and we get 2. This is mod arithmetic. Given
a nonnegative integer n and a positive integer p, the mod function
is defined as follows: "n mod p" is the integer remainder obtained
when n is divided by p. Symbolically, we write n mod $p = r$ if and
only if $n = pq + r$, where q and r are integers with the property of
$0 \leq r < p$. Further, if n mod $p = 0$, then p divides into n exactly, with
no remainder. So, for example, 26 mod 24 = 2, since $26 = 24 \times 1 + 2$
(we read this as 24 divides into 26 once, with a remainder of 2).

Back to the Josephus problem. Let p designate the position of the survivor. For n beginning at 6, we rewrite the previous table to get

n	p
6	1
7	4
8	7
9	$10 \rightarrow 1$

Since 10 mod 9 = 1, the survivor position p becomes 1. Continue adding 3 to p as n increases by 1. For $n = 10$, $p = 4$; for $n = 11$, $p = 7$. Continue to add 3 until $p \geq n$, when we replace p with p mod n. When $n = 14$ men, we have $p = 16 > 14$, so 16 mod 14 = 2, and the actual position is 2. Therefore, when $n = 14$, $p = 2$. Continue adding 3 to p until we reach $n = 21$ men, and we find $p = 23$ mod 21 = 2. With $n = 31$, $p = 32$ mod 31 = 1, and with $n = 41$ men, we end in position 31. Did you follow the algorithm?

By knowing one solution and performing modulo addition, we can iterate to the survivor position for any n. We keep adding 3 to the survivor position as long as $p + x \times 3 \leq n + x$, where x is the number of men greater than n. As soon as $p + x \times 3 > n + x$, we start with a new p, n and begin with $x = 1$ again. In Figure 9.2, we show the survivor positions as the number of men in the circle increases. You can see the parallel lines, which all increase by 3 as the recursion continues until they reach the point where $p + x \times 3 > n + x$.

What about the next-to-last survivor? Recall that Hercules will spare two men. Look again at the $n = 6$ situation. The survivor is in position 1. So, when $n = 7$, the sixth (next-to-last) man removed is in position 1! Let q be the next-to-last man removed, and we iterate as before, keeping track of our addition:

n	q
7	1
8	4
9	7
10	10
11	$13 \rightarrow 2$

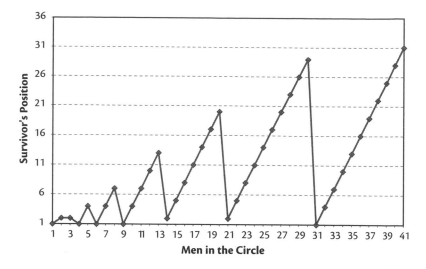

Figure 9.2. The survivor position as a function of the number of men in the circle.

The new q, associated with $n = 11$, is 2, and the iteration continues. When there are 41 men, $q = 16$. Check to make sure. So, given 41 men that have been captured, Hercules will spare those in positions 16 and 31 after killing every third man.

This problem is made easy with a spreadsheet, or you can work it online with a number of Flash or Java applets. Enter values for n and k (how many men are skipped each time) and click "Start." Try to guess the answer using the above approach before you watch the applet give it to you.

In *Concrete Mathematics*, Graham, Knuth, and Patashnik developed a recursive solution to the Josephus problem where every third person is removed. They define $J_k(n)$ to be the survivor's position number. In particular, $J_3(n)$ is the survivor's number with every third man removed. They offer the following recurrence relationship for any n:

$$J_3(n) = \lceil \tfrac{3}{2} J_3 \left(\lfloor \tfrac{2}{3}n \rfloor \right) + a_n \rceil \bmod n + 1, \qquad (9.1)$$

with $J_3(6) = 1$ as an initial condition.

Here the parameter a_n is defined as -2, $+1$, or $-\frac{1}{2}$ according to whether $n \bmod 3 = 0$, 1, or 2, respectively. $\lceil\,\rceil$ designates the ceiling, and $\lfloor\,\rfloor$ designates the floor. The floor of a number is the greatest integer less than or equal to the number, and the ceiling of a number is the least integer greater than or equal to the number, unless the number itself is already an integer (in which case the floor and ceiling equal the number). Symbolically, given a number x, $z \leq \lfloor x \rfloor < z+1$ and $z - 1 < \lceil x \rceil \leq z$, where z is an integer.

Graham, Knuth, and Patashnik realize the merit and difficulty of Equation 9.1, writing, "this recurrence is too horrible to pursue." But it works! For example, suppose $n = 10$. Then

$$J_3(10) = \lceil \frac{3}{2} J_3 \left(\lfloor \frac{2}{3} \times 10 \rfloor \right) + a_n \rceil \bmod 10 + 1.$$

First, since $10 \bmod 3 = 1$, $a_n = +1$. We also need to know that $J_3(6) = 1$. Then,

$$J_3(10) = \lceil \frac{3}{2} J_3 \left(\lfloor \frac{20}{3} \rfloor \right) + 1 \rceil \bmod 10 + 1$$

$$= \lceil \frac{3}{2} J_3(6) + 1 \rceil \bmod 10 + 1$$

$$= \lceil \frac{3}{2} \times 1 + 1 \rceil \bmod 10 + 1$$

$$= \lceil \frac{5}{2} \rceil \bmod 10 + 1$$

$$= 3 \bmod 10 + 1$$

$$= 3 + 1$$

$$= 4,$$

which is what we obtained earlier.

When we use the Graham-Knuth-Patashnik formula for $n = 41$, we must also know $J_3(27)$, which is 20, and the relation leads to the desired effect, $J_3(41) = 31$. However, Hercules would agree with the authors of *Concrete Mathematics* in that the proof of this formula is indeed too horrible.

This type of problem also attracted the attention of Leonhard Euler, who wrote a paper in Latin titled *Observationes circa novum et singulare progressionum genus* in 1776. The title translates to "Observations on a new and singular type of progression." According to the Euler Archives, this work (E476) was presented to St. Petersburg Academy on July 4, 1771, and was published 5 years later. Euler studied the similar problem of 30 men in a boat, 15 of whom are Christians and 15 of whom are pagans. The boat has encountered a dreadful storm, and the captain fears that all will perish unless half the crew is thrown overboard. Who should be sacrificed, Euler ponders. Again the 30 men are placed in a circle, and every ninth man is thrown overboard. Where should the Christians be placed so that all 15 pagans are thrown overboard? Euler studies the "sequence of castoffs," in various specific cases and offers interesting properties for each case.

9.2.2. The Amazons and the Spread of a Rumor

From Apollodorus:

Having put in at the harbor of Themiscyra, Hercules received a visit from Hippolyte, who inquired why he was come, and promised to give him the belt. But Hera in the likeness of an Amazon went up and down the multitude saying that the strangers who had arrived were carrying off the queen. So the Amazons in arms charged on horseback down on the ship.

TASK: Hera, the wife of Zeus who wants Hercules to fail, disguises herself as an Amazon warrior. In this likeness, she travels "up and down the multitude" of the other Amazons, spreading the false rumor that Hercules and his band of strangers are carrying off the Amazon queen. Within 5 minutes, each Amazon warrior who hears the story passes it on to two other warriors who have not yet heard the rumor. Hera continues to spread the rumor as she passes down the ranks of the Amazons, telling a group of four Amazon women every 5 minutes. Assume she tells only four different groups of warriors and then disappears. How long will it take before 1500 warriors have heard the lie?

SOLUTION: In this task, we need to model the number of Amazon warriors who have been told the rumor that the queen is being carried off by Hercules and his crew. Let's define $w(t)$ as the number of Amazon warriors who have been told at time t. Since we are given that it takes 5 minutes for the rumor to multiply, let's define a time interval to be 5 minutes. This means that $w(1)$ is the number of warriors who have been told after 5 minutes, $w(4)$ is the number of warriors who have been told after 4 times 5 (or 20) minutes, and so forth. We need to determine the value of t when $w(t) = 1500$.

Let's enumerate the first few time intervals. Initially, no one has been told anything, so $w(0) = 0$. After 5 minutes, Hera has told 4 warriors, so $w(1) = 4$. After 10 minutes, each of these 4 warriors has told 2 others, and Hera has additionally told 4 new warriors. Therefore, $w(2)$ is equal to $w(1)$ plus those newly told, or

$$w(2) = w(1) + 2\, w(1) + 4 = 16.$$

Continuing in this fashion,

$$w(3) - 3\, w(2) + 4 - 52,$$

and so forth. In general, the recursive relationship is

$$w(t + 1) = 3\, w(t) + 4 \tag{9.2}$$

for $t = 0$, 1, 2, 3, and 4. After $t = 4$, Hera ceases to tell any more warriors (and $w(4) = 160$). From this time on, the number can be predicted with a geometric sequence. The new model is simply

$$w(t + 1) = 3\, w(t), \qquad t \geq 4. \tag{9.3}$$

Equation 9.3 has the solution $w(k) = b\,(3)^k$ for some future value $k \geq 4$ and some constant b. The new initial condition is $w(4) = 160$, so

$$w(4) = b\,(3)^4 = 160,$$

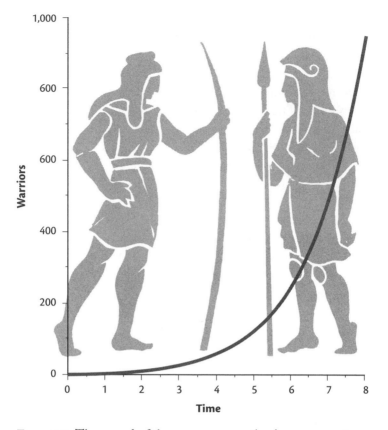

Figure 9.3. The spread of the rumor among the Amazon warriors.

which gives $b = \frac{160}{81}$. The particular solution is then $w(k) = \frac{160}{81}(3)^k$. Solving now for the time required to reach 1500, we obtain

$$ k = \frac{\ln \frac{1500 \times 81}{160}}{\ln 3} = 6.03 \approx 6. $$

This tells us that after about $6 \times 5 = 30$ minutes, 20 minutes with Hera involved and 10 minutes without Hera, 1500 warriors are looking for Hercules. We can see the exponential growth of the spread of the rumor in Figure 9.3.

Figure 9.3 may be misleading. To show the growth of the rumor, I have "smoothed" the curve, connecting the discrete values in a

continuous curve. Since this is a discrete equation, we should not try
to extrapolate to an *exact* time. Since each time interval is 5 minutes,
we should not try to predict within the interval. The best way to state
the solution is that some time between 30 and 35 minutes (since $6 \leq$
$6.03 \leq 7$), 1500 Amazon warriors will have been told that Hercules is
abducting their queen.

9.2.3. *Exercise: Hercules and the Kraken*

From Apollodorus:

> But it chanced that the city was then in distress consequently on the wrath
> of Apollo and Poseidon. For desiring to put the wantonness of Laomedon to
> the proof, Apollo and Poseidon assumed the likeness of men and undertook
> to fortify Pergamum for wages. But when they had fortified it, he would not
> pay them their wages. Therefore Apollo sent a pestilence, and Poseidon a
> sea monster, which, carried up by a flood, snatched away the people of the
> plain. But as oracles foretold deliverance from these calamities if Laomedon
> would expose his daughter Hesione to be devoured by the sea monster, he
> exposed her by fastening her to the rocks near the sea. Seeing her exposed,
> Hercules promised to save her on condition of receiving from Laomedon the
> mares which Zeus had given in compensation for the rape of Ganymede. On
> Laomedon's saying that he would give them, Hercules killed the monster
> and saved Hesione.

TASK: The Kraken was a terrifying sea monster who instilled fear in
the ancient Greeks because of its size and appearance. It climbs onto
the rocks at the edge of the shore and raises itself up before turning its
attention to the princess Hesione, who is tied to the rocks. In order to
instill the most fear, the Kraken ensures that its height is at a position
where the angle of sight from the people is at a maximum. If the Kraken
stands 38 meters high and climbs onto the rocks, which rise 17 meters
above sea level, how far away do the villagers need to be to get the best
(most fearful) view of the monster (see Figure 9.1)?

SOLUTION: This problem is adapted from an optimization problem
found in Hughes-Hallett's calculus text (see the Bibliography). The
story of Hercules saving Hesione is similar to that of Perseus

and Andromeda (made famous in the movie *Clash of the Titans*). Apollodorus translator Sir James George Frazer believes that "both tales may have originated in a custom of sacrificing maidens to be the brides of the Sea." Before Hercules comes to the rescue, however, we must find the best vantage point for the Greeks to watch the spectacle. The Kraken was the last of the Titans, a race of tremendous monsters that predated the Greek gods. If the sacrifice of a maiden was not made, Poseidon released the Kraken, a gigantic swimming creature with four arms, to destroy a city by flooding the entire kingdom.

Now, we turn to solving the task. Let θ be the angle between the top of the rocks and the top of the Kraken's head and let s be the horizontal distance from the villagers to the rocks, assuming they watch the scene from sea level. The task is to maximize the angle θ, which is measured from the top of the rocks to the top of the Kraken. Let's split θ into two angles, defining $\theta = \theta_1 - \theta_2$. We will use θ_1 to measure the angle between the base of the rocks (sea level) and the top of the Kraken, and θ_2 will be the angle from the base of the rocks to the top of the rocks. Knowing the height of the rocks and the Kraken, we can write

$$\tan(\theta_1) = \frac{55}{s} \quad \text{and} \quad \tan(\theta_2) = \frac{17}{s}.$$

Solving for the two angles, we get

$$\theta_1 = \arctan\left(\frac{55}{s}\right),$$

$$\theta_2 = \arctan\left(\frac{17}{s}\right),$$

so

$$\theta = \arctan\left(\frac{55}{s}\right) - \arctan\left(\frac{17}{s}\right), \tag{9.4}$$

which is valid for all values of $s > 0$. The task is to now find all the critical points of Equation 9.4. We do this by finding the first derivative using the chain rule:

$$\frac{d\theta}{ds} = \frac{1}{1 + \left(\frac{55}{s}\right)^2}\left(\frac{-55}{s^2}\right) - \frac{1}{1 + \left(\frac{17}{s}\right)^2}\left(\frac{-17}{s^2}\right)$$

$$= \frac{-55}{s^2 + 55^2} + \frac{17}{s^2 + 17^2}$$

$$= \frac{-38\,(s^2 - 935)}{(s^2 + 55^2)\,(s^2 + 17^2)}. \tag{9.5}$$

To find the critical points, we solve $\frac{d\theta}{ds} = 0$. Setting the numerator equal to zero, we can obtain $s^2 - 935 = 0$, or $s \approx \pm 30.58$ meters. We are concerned only with the positive solution (Equation 9.4 is defined for $s > 0$). We need to verify that this is a maximum. We see that for $0 < s < 30.58$, the first derivative is positive, yet for $s > 30.58$, the first derivative is negative. Thus, by the First Derivative Test, we can say that the angle θ has a maximum at $s = 30.58$ meters.

We substitute this value of s into Equation 9.4. The value of θ is then approximately $31.85°$. Alternatively, we could calculate the second derivative (a bit more complicated) and plot it. In doing so, we would see that the second derivative is everywhere negative in a neighborhood about the critical point, which implies that the function is concave down, indicating that $s \approx 30.58$ meters is a maximum. The distance of 30.58 meters is still pretty close for the villagers to watch Hesione being sacrificed to the Kraken, but the Greeks always loved a good show.

The "Second Vatican Mythographer" wrote

Hercules, in full armour, leaped into the jaws of the sea-monster, and was in its belly for three days hewing and hacking it, and that at the end of the three days, he came forth without any hair on his head.

The Vatican mythographers provide a sort of source book about Greek and Roman myth. Apollodorus does not elaborate on how Hercules defeats the Kraken. After doing so, however, our hero brings the Belt of Hippolyte back to Mycenae and accepts the next labor.

The Tenth Labor: Geryon's Cattle

From Apollodorus:

As a tenth labour he was ordered to fetch the kine of Geryon from Erythia. Now Erythia was an island near the ocean; it is now called Gadira. This island was inhabited by Geryon, son of Chrysaor by Callirrhoe, daughter of Ocean. He had the body of three men grown together and joined in one at the waist, but parted in three from the flanks and thighs. He owned red kine, of which Eurytion was the herdsman and Orthus, the two-headed hound, begotten by Typhon on Echidna, was the watchdog. So journeying through Europe to fetch the kine of Geryon he destroyed many wild beasts and set foot in Libya, and proceeding to Tartessus he erected as tokens of his journey two pillars over against each other at the boundaries of Europe and Libya. But being heated by the Sun on his journey, he bent his bow at the god, who in admiration of his hardihood, gave him a golden goblet in which he crossed the ocean. And having reached Erythia he lodged on Mount Abas. However the dog, perceiving him, rushed at him; but he smote it with his club, and when the herdsman Eurytion came to the help of the dog, Hercules killed him also. But Menoetes, who was there pasturing the kine of Hades, reported to Geryon what had occurred, and he, coming up with Hercules beside the river Anthemus, as he was driving away the kine, joined battle with him and was shot dead. And Hercules, embarking the kine in the goblet and sailing across to Tartessus, gave back the goblet to the Sun.

And passing through Abderia he came to Liguria, where Ialebion and Dercynus, sons of Poseidon, attempted to rob him of the kine, but he killed them and went on his way through Tyrrhenia. But at Rhegium a bull broke away and hastily plunging into the sea swam across to Sicily, and having passed through the neighboring country since called Italy after it, for the Tyrrhenians called the bull italus, came to the plain of Eryx, who reigned over the Elymi. Now Eryx was a son of Poseidon, and he mingled the bull with his own herds. So Hercules entrusted the kine to Hephaestus and hurried away in search of the bull. He found it in the herds of Eryx, and when the king refused to surrender it unless Hercules should beat him in a wrestling bout, Hercules beat him thrice, killed him in the wrestling, and taking the bull drove it with the rest of the herd to the Ionian Sea. But when he came to the creeks of the sea, Hera afflicted the cows with a gadfly, and they dispersed among the skirts of the mountains of Thrace. Hercules went in pursuit, and having caught some, drove them to the Hellespont; but the remainder were thenceforth wild. Having with difficulty collected the cows, Hercules blamed the river Strymon, and whereas it had been navigable before, he made it unnavigable by filling it with rocks; and he conveyed the kine and gave them to Eurystheus, who sacrificed them to Hera.

10.1. The Tasks

Geryon was a monster who had three bodies joined at the waist. He lived on the island of Erithia with his two-headed dog Orthus, and he possessed a wonderful herd of red cattle. The mighty Hercules has four tasks to complete. First, he must determine the optimum light intensity between the lighthouses situated on **The Pillars of Hercules**. For the second task, Hercules receives a goblet from the Sun and must maneuver it through the Strait of Gibraltar, dealing with a current of 0.4 meter per second, in **The Golden Goblet** problem. Once Hercules has driven the cattle to the Ionian Sea, Hera sends a plague of gadflies into the herd, and Hercules must model the spread of gadfly disease, in the **Hera Sends the Gadflies** task. For a fourth and final task, Hercules is determined to make the River Strymon unnavigable. Help him to determine the highest packing factor of the rocks in the **Blocking the River Strymon** problem.

CHAPTER 10

10.1.1. The Pillars of Hercules

TASK: Hercules erects "two pillars over against each other at the boundaries of Europe and Libya." Today, these are known as the Pillars of Hercules, consisting of Mons Calpe in Europe (the Rock of Gibraltar) and Mons Abyla in Africa. The Sun places a lighthouse on each pillar, and Hercules journeys across the Strait of Gibraltar in a golden goblet in a straight line from one pillar to the other. The first lighthouse (on Gibraltar) reaches a height of 426 meters above sea level, the second, 204 meters. The two pillars with lighthouses are 22.7 kilometers apart. At what distance from Gibraltar will the angles from each lighthouse to the goblet be the same? Further, the intensity $I(x)$ of the light from the lighthouses is inversely proportional to the square of the distance from each lighthouse. Suppose that the combined light intensity on a straight line, at a distance x from Gibraltar, is given by $I(x) = 5/x^2 + 1/(22700 - x)^2$. Find the point on the line between Mons Calpe and Mons Abyla where the intensity of the light is at a minimum.

10.1.2. The Golden Goblet

TASK: As stated in the first task, the distance between the Pillars of Hercules is 22,700 meters. The average current flows through the Strait of Gibraltar at approximately 0.4 meter per second (perpendicular to the pillars). Assuming Hercules can power the goblet to move through the water at 2.75 meters per second, at what angle relative to his departure point must he steer in order to reach Africa from Gibraltar? How long will the trip take?

10.1.3. Hera Sends the Gadflies

TASK: Hera sends gadflies to afflict Geryon's cattle. The gadflies spread a nonfatal disease through the kine population. This disease causes them to disperse throughout the mountains of Thrace. Suppose there are 250 cows. Healthy cattle are infected by the gadflies at the rate of a certain number of cattle per day, which is proportional to the product of the number of infected cattle and the number of uninfected (or healthy) cattle, divided by the total population. The rate of recovery per day is 80 percent of the number of infected cattle. Model the spread of gadfly disease. If 20 cows are initially infected, what percentage of the population will ultimately get the disease?

10.1.4. Blocking the River Strymon

TASK: Hercules wants to make the River Strymon unnavigable. To do so, he will fill it with rocks. Assuming he uses spherically shaped rocks which are all roughly the same size, he sets the rocks in the riverbed to form lattices. Help Hercules determine which lattice has the highest packing factor. Simply put, a lattice is a three-dimensional repeating pattern. Hercules must choose between the following types: simple cubic (sc), body-centered cubic (bcc), and face-centered cubic (fcc). An sc lattice has a rock only at each corner of the cube. A bcc lattice has a rock at each corner and another at the body center of the cube. An fcc lattice has a rock at each corner and one at each face of the cube but not at the center. See Figure 10.5 for a view of each type. The packing factor is the volume fraction of the basic unit cell (in this case, the cube) that is occupied by the spherical rocks.

10.2. The Solutions

10.2.1. The Pillars of Hercules

From Apollodorus:

So journeying through Europe to fetch the kine of Geryon he destroyed many wild beasts and set foot in Libya, and proceeding to Tartessus he erected as tokens of his journey two pillars over against each other at the boundaries of Europe and Libya. But being heated by the Sun on his journey, he bent his bow at the god, who in admiration of his hardihood, gave him a golden goblet in which he crossed the ocean.

TASK: Hercules erects "two pillars over against each other at the boundaries of Europe and Libya." Today, these are known as the Pillars of Hercules, consisting of Mons Calpe in Europe (the Rock

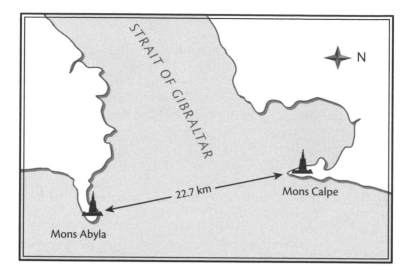

Figure 10.1. The Pillars of Hercules.

of Gibraltar) and Mons Abyla in Africa. The Sun places a lighthouse on each pillar, and Hercules journeys across the Strait of Gibraltar in a golden goblet in a straight line from one pillar to the other. The first lighthouse (on Gibraltar) reaches a height of 426 meters above sea level, the second, 204 meters. The two pillars with lighthouses are 22.7 kilometers apart. At what distance from Gibraltar will the angles from each lighthouse to the goblet be the same? Further, the intensity $I(x)$ of the light from the lighthouses is inversely proportional to the square of the distance from each lighthouse. Suppose that the combined light intensity on a straight line, at a distance x from Gibraltar, is given by $I(x) = 5/x^2 + 1/(22700 - x)^2$. Find the point on the line between Mons Calpe and Mons Abyla where the intensity of the light is at a minimum.

SOLUTION: Refer to Figure 10.1. Although at its narrowest point the Strait of Gibraltar is about 13 kilometers wide, it is 22.7 kilometers from one pillar to the other. Let x be the distance from Mons Calpe to the goblet and let y be the distance from Mons Abyla to the goblet. We seek to determine the value of x when the values of θ and ϕ will be the same.

From Figure 10.1, we can use similar triangles to find x when $\tan(\theta) = \tan(\phi)$. Or, more specifically, we need to find x such that

$$\tan(\theta) = \frac{22700 - x}{219} = \frac{x}{441} = \tan(\phi).$$

This yields $x \approx 15{,}167.7$ meters, and thus $y = 22700 - x \approx 7532.3$ meters. It appears that the two angles are equal ($\approx 88.33°$—this corresponds to an angle of elevation of $\approx 1.67°$) when Hercules in the golden goblet is just past the two-thirds point in his journey across the Strait of Gibraltar.

The Europa Lighthouse really does sit on the Gibraltar peninsula, although it is not at the top of the Rock. Mons Abyla is also located on a peninsula and is topped by a fort, the Fortaleza de Hacho, which was first built by the Byzantines. The Rock of Gibraltar rises 426 meters up from the sea, and Mons Abyla rises 204 meters. We are given that the light intensity is modeled by the equation

$$I(x) = \frac{5}{x^2} + \frac{1}{(22700 - x)^2}. \tag{10.1}$$

The derivative is

$$I'(x) = \frac{-10}{x^3} + \frac{2}{(22700 - x)^3},$$

which we can set equal to zero to find the critical points x. This simplifies to solving $-10(22700 - x)^3 + 2x^3 = 0$, or even simpler, $5(22700 - x)^3 = x^3$. Taking cube roots and solving for x yields the critical value $x = 14{,}323.5$ meters, or about 14.3 kilometers.

To verify that this x minimizes $I(x)$, we take the second derivative:

$$I''(x) = \frac{30}{x^4} + \frac{6}{(22700 - x)^4}. \tag{10.2}$$

The right-hand side of Equation 10.2 will be positive for any value of x in the domain $0 < x < 22700$, so we indeed have found the minimum value. In addition, a plot of $f(x)$ is shown in Figure 10.2. You can see the minimum value at $x \approx 14{,}323.5$ meters. Therefore, the light intensity between the Pillars of Hercules will be at a minimum when the goblet is 14,323.5 meters from Mons Calpe on a straight line to

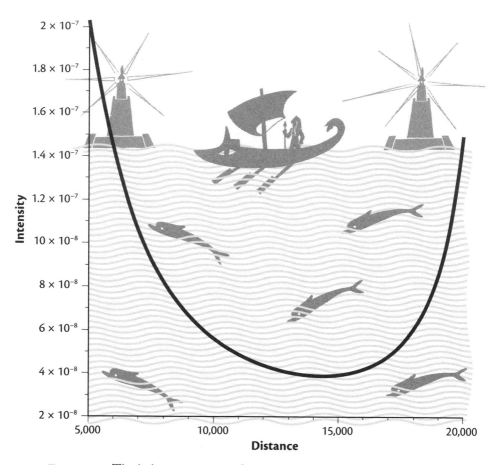

Figure 10.2. The light intensity as a function of the distance from Gibraltar.

Mons Abyla. In his book *Timaeus*, Plato wrote that the lost realm of Atlantis was situated beyond the Pillars of Hercules. Does this mean that Hercules could have traveled to Atlantis? More myths....

10.2.2. The Golden Goblet

From Apollodorus:

But being heated by the Sun on his journey, he bent his bow at the god, who in admiration of his hardihood, gave him a golden goblet in which he crossed the ocean.

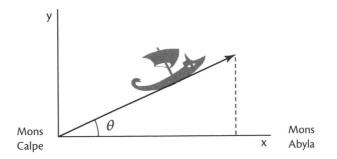

Figure 10.3. Traveling in the golden goblet.

TASK: As stated in the first task, the distance between the Pillars of Hercules is 22,700 meters. The average current flows through the Strait of Gibraltar at approximately 0.4 meter per second (perpendicular to the pillars). Assuming Hercules can power the goblet to move through the water at 2.75 meters per second, at what angle relative to his departure point must he steer in order to reach Africa from Gibraltar? How long will the trip take?

SOLUTION: Now, let's look at the path of the golden goblet given to Hercules by the Sun (see Figure 10.3). First, we make some assumptions. Let's assume that the two given speeds (that of the current and that of the boat) remain constant during the duration of travel. Also, we will assume that once Hercules sets his initial heading, based on the required angle, he does not deviate from that bearing (that is, the angle his boat makes with the line perpendicular to the strait remains constant). As the boat travels upstream, the current pushes it back downstream.

This now becomes a geometry/trigonometry problem. If there was no current associated with the Strait of Gibraltar (from the Atlantic Ocean into the Mediterranean Sea), Hercules could travel directly from one pillar to the other in a straight line and the time required would be equal to the distance (22.7 kilometers) divided by the boat's crossing speed (2.75 meters per second). Taking the current into account does not pose any difficulties, though. With each second, the golden goblet travels 2.75 meters upstream. The horizontal distance traveled (denoted by x) equals the projection of the heading in the x-direction, or

$$x = 2.75 \ \cos \theta,$$

where θ is the heading. Thus, in t seconds, $x = (2.75 \ \cos \theta)t$. Similarly, in the absence of a current, the vertical distance traveled (y) equals

$$y = 2.75 \ \sin \theta.$$

If Hercules were to steer his boat directly toward the second pillar at Mons Abyla and not adjust his heading, the current of the water through the Strait of Gibraltar would cause the boat to drift 0.4 meter per second. Dividing the drift speed by the boat's speed and taking the arcsine, we find that the boat would actually drift 3337.3 meters from its destination. However, we have to account for the current, subtracting it as the boat moves upstream, and after t seconds, we find that $y = (2.75 \sin \theta - 0.4)t$.

For the golden goblet to land at Mons Abyla in Africa, the value of y must be zero. Therefore, the critical angle satisfies

$$(2.75 \sin \theta - 0.4)t = 0,$$

which tells us that

$$\theta = \sin^{-1}\left(\frac{0.4}{2.75}\right) \approx 0.146 \text{ radian}.$$

This corresponds to an angle of roughly 8.36°. The time required to reach the second pillar is found by taking the horizontal distance and dividing it by the speed of the goblet, or

$$t = \frac{22700}{2.75 \ \cos(0.146)} \approx 8343.28 \text{ seconds}.$$

This is just over 2 hours and 19 minutes.

10.2.3. Hera Sends the Gadflies

From Apollodorus:

Hercules, ... taking the bull drove it with the rest of the herd to the Ionian Sea. But when he came to the creeks of the sea, Hera afflicted the cows

with a gadfly, and they dispersed among the skirts of the mountains of
Thrace. Hercules went in pursuit, and having caught some, drove them to
the Hellespont; but the remainder were thenceforth wild.

TASK: Hera sends gadflies to afflict Geryon's cattle. The gadflies spread
a nonfatal disease through the kine population. This disease causes
them to disperse throughout the mountains of Thrace. Suppose there
are 250 cows. Healthy cattle are infected by the gadflies at the rate of a
certain number of cattle per day, which is proportional to the product of
the number of infected cattle and the number of uninfected (or healthy)
cattle, divided by the total population. The rate of recovery per day
is 80 percent of the number of infected cattle. Model the spread of
gadfly disease. If 20 cows are initially infected, what percentage of the
population will ultimately get the disease?

SOLUTION: Hera has a history of using gadflies in Greek mythology. In
another story, she torments Io, the heifer maiden. Hera's husband Zeus
lusts after Io and eventually turns her into a white heifer to hide her
from Hera, a jealous wife. Hera is not fooled and demands Io as a gift
from Zeus. She then sends Argus, the 100-eyed monster, to guard Io
and later sends gadflies to torment and sting Io, forcing her to wander
farther and farther away from home.

Back to our problem with Geryon's cattle. Let $I(t)$ designate the
number of infected cattle on day t. Then the number of healthy
cattle is $250 - I$. The rate of infection, given in the task, is written
symbolically as

$$\text{infection rate} = \frac{1}{250}(250 - I)I. \qquad (10.3)$$

We are given the rate of recovery:

$$\text{recovery rate} = 0.8I. \qquad (10.4)$$

The rate of change in the number of infected cattle per day is
equal to the infection rate minus the recovery rate. So, we subtract

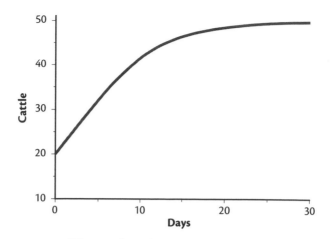

Figure 10.4. The number of cattle infected by gadfly disease.

Equation 10.4 from Equation 10.3, obtaining

$$\frac{dI}{dt} = \frac{1}{250}(250 - I)I - 0.8I,$$

$$= I - \frac{I^2}{250} - 0.8I,$$

$$= 0.2I\left(1 - \frac{I}{50}\right).$$

This is a logistic differential equation, with $I(0) = 20$. A plot of the solution is given in Figure 10.4. The equilibrium value for the number of infected cattle is found by solving $\frac{dI}{dt} = 0$. Doing so yields either $I(t) = 0$ (the trivial equilibrium value) or $I(t) = 50$. As t gets large, $I(t) \to 50$. This is verified in the figure. In the long run, only 50 cattle (about 20 percent of the total population) will be infected on any day. These are probably the ones who disperse into the mountains.

10.2.4. Blocking the River Strymon

From Apollodorus:

Having with difficulty collected the cows, Hercules blamed the river Strymon, and whereas it had been navigable before, he made it unnavigable

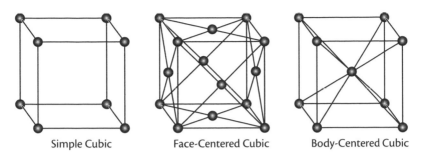

| Simple Cubic | Face-Centered Cubic | Body-Centered Cubic |

Figure 10.5. The three cubic lattice structures.

by filling it with rocks; and he conveyed the kine and gave them to Eurystheus, who sacrificed them to Hera.

TASK: Hercules wants to make the River Strymon unnavigable. To do so, he will fill it with rocks. Assuming he uses spherically shaped rocks which are all roughly the same size, he sets the rocks in the riverbed to form lattices. Help Hercules determine which lattice has the highest packing factor. Simply put, a lattice is a three-dimensional repeating pattern. Hercules must choose between the following types: simple cubic (sc), body-centered cubic (bcc), and face-centered cubic (fcc). An sc lattice has a rock only at each corner of the cube. A bcc lattice has a rock at each corner and another at the body center of the cube. An fcc lattice has a rock at each corner and one at each face of the cube but not at the center. See Figure 10.5 for a view of each type. The packing factor is the volume fraction of the basic unit cell (in this case, the cube) that is occupied by the spherical rocks.

SOLUTION: Most metals, ceramics, and polymers have an underlying structure that is ordered, repeating, and three-dimensional, the properties of a lattice. The atoms in these structures form *crystals*, and the crystals are composed of lattices in which the atoms are arranged. Materials scientists study such ordered and repeating structures, and there are seven different possible crystal systems which differ by coordinate axes lengths and axial angles (the angles between the coordinate axes). The seven systems are cubic, tetragonal, orthorhombic, monoclinic, triclinic, hexagonal, and rhombohedral.

Hercules' task is concerned with the first crystal system, the cubic, where the axial length in each direction is the same and all the angles between axes are 90° (hence forming a cube).

Hercules must compare the packing factors of simple cubic, body-centered cubic, and face-centered cubic lattices. A simple cubic unit cell, which is the building block of the lattice, has atoms at each corner of the cube. We will assume that all rocks are spheres and discuss the locations of these spheres inside the unit cube. In this task, Hercules places the rocks in such a manner that they stack up in a series of cubes. At each corner of the cube is $\frac{1}{8}$ of a sphere. Since there are 8 corners in a cube, the simple cubic lattice has 1 full sphere in each unit cell. The packing factor (PF) is the ratio of the volume of the spheres to the total volume of the cube. Let r denote the radius of each sphere. Then the packing factor is

$$PF = \frac{\frac{4}{3}\pi r^3}{(2r)^3} = \frac{\pi}{6} \approx 52.4\,\text{percent.}$$

This means that a little more than half of the unit cell for a simple cubic lattice is filled by the rocks, leaving plenty of space between the rocks.

Next we investigate the bcc lattice. The bcc unit cell has a sphere at each corner and another at the cube's center. This means that there are 2 full spheres in each cell. Since there is a sphere in the center, the spheres at each corner do not touch as in the simple cubic lattice. We can relate the radius of each sphere r to the length of one side of the cube (which we will call a). Across the cube diagonal, the length is $4r$ since the corner spheres touch the center sphere. So, keeping in mind the 2 complete spheres per unit cell,

$$PF = \frac{2\left(\frac{4}{3}\pi r^3\right)}{a^3} = \frac{\frac{8}{3}\pi r^3}{\left(\frac{4r}{\sqrt{3}}\right)^3} = \frac{\sqrt{3}\pi}{8} \approx 68.0\,\text{percent.}$$

The fcc lattice has 4 spheres per unit cell since there are spheres at each corner and on each face. Adding the $\frac{1}{8}$ spheres times 8 corners and the $\frac{1}{2}$ spheres times 6 faces, we find that there is a total of 4 spheres

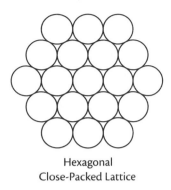

Hexagonal
Close-Packed Lattice

Figure 10.6. The hexagonal close-packed lattice structure.

per unit cell. The cube length a is related to the sphere radius by

$$a = \frac{4r}{\sqrt{2}},$$

so the packing factor is

$$PF = \frac{4\left(\frac{4}{3}\pi r^3\right)}{a^3} = \frac{\frac{16}{3}\pi r^3}{\left(\frac{4r}{\sqrt{2}}\right)^3} = \frac{\sqrt{2}\pi}{6} \approx 74.0 \text{ percent.}$$

So, of the three considered, the fcc lattice has the highest packing factor. Realistically, it would take a great effort to align the rocks in the river in any one of these three lattices. A more realistic approach would involve the hexagonal close-packed (hcp) lattice, which is shown in Figure 10.6. The packing factor of an hcp lattice is identical to that of an fcc lattice, 74 percent. The hcp lattice is typically seen in grocery stores in the stacking of oranges and other round fruits and vegetables.

CHAPTER  11

The Eleventh Labor: The Apples of the Hesperides

From Apollodorus:

When the labours had been performed in eight years and a month, Eurystheus ordered Hercules, as an eleventh labour, to fetch golden apples from the Hesperides, for he did not acknowledge the labour of the cattle of Augeas nor that of the hydra. These apples were not, as some have said, in Libya, but on Atlas among the Hyperboreans. They were presented by Earth to Zeus after his marriage with Hera, and guarded by an immortal dragon with a hundred heads, offspring of Typhon and Echidna, which spoke with many and divers sorts of voices. With it the Hesperides also were on guard, to wit, Aegle, Erythia, Hesperia, and Arethusa. So journeying he came to the river Echedorus. And Cycnus, son of Ares and Pyrene, challenged him to single combat. Ares championed the cause of Cycnus and marshalled the combat, but a thunderbolt was hurled between the two and parted the combatants. And going on foot through Illyria and hastening to the river Eridanus he came to the nymphs, the daughters of Zeus and Themis. They revealed Nereus to him, and Hercules seized him while he slept, and though the god turned himself into all kinds of shapes, the hero bound him and did not release him till he had learned from him where were the apples and the Hesperides. Being informed, he traversed Libya. That country was then ruled by Antaeus, son of Poseidon, who used to kill strangers by forcing them to wrestle. Being forced to wrestle with him, Hercules hugged him, lifted him aloft, broke and killed him; for when he touched earth so it was that he waxed stronger, wherefore some said that he was a son of Earth.

After Libya he traversed Egypt. That country was then ruled by Busiris, a son of Poseidon by Lysianassa, daughter of Epaphus. This Busiris used to sacrifice strangers on an altar of Zeus in accordance with a certain oracle. For Egypt was visited with dearth for nine years, and Phrasius, a learned seer who had come from Cyprus, said that the dearth would cease if they slaughtered a stranger man in honor of Zeus every year. Busiris began by slaughtering the seer himself and continued to slaughter the strangers who landed. So Hercules also was seized and haled to the altars, but he burst his bonds and slew both Busiris and his son Amphidamas.

And traversing Asia he put in to Thermydrae, the harbor of the Lindians. And having loosed one of the bullocks from the cart of a cowherd, he sacrificed it and feasted. But the cowherd, unable to protect himself, stood on a certain mountain and cursed. Wherefore to this day, when they sacrifice to Hercules, they do it with curses.

And passing by Arabia he slew Emathion, son of Tithonus, and journeying through Libya to the outer sea he received the goblet from the Sun. And having crossed to the opposite mainland he shot on the Caucasus the eagle, offspring of Echidna and Typhon, that was devouring the liver of Prometheus, and he released Prometheus, after choosing for himself the bond of olive, and to Zeus he presented Chiron, who, though immortal, consented to die in his stead. Now Prometheus had told Hercules not to go himself after the apples but to send Atlas, first relieving him of the burden of the sphere; so when he was come to Atlas in the land of the Hyperboreans, he took the advice and relieved Atlas. But when Atlas had received three apples from the Hesperides, he came to Hercules, and not wishing to support the sphere he said that he would himself carry the apples to Eurystheus, and bade Hercules hold up the sky in his stead. Hercules promised to do so, but succeeded by craft in putting it on Atlas instead. For at the advice of Prometheus he begged Atlas to hold up the sky till he should put a pad on his head. When Atlas heard that, he laid the apples down on the ground and took the sphere from Hercules. And so Hercules picked up the apples and departed. But some say that he did not get them from Atlas, but that he plucked the apples himself after killing the guardian snake. And having brought the apples he gave them to Eurystheus. But he, on receiving them, bestowed them on Hercules, from whom Athena got them and conveyed them back again; for it was not lawful that they should be laid down anywhere.

11.1. The Tasks

The Hesperides were known as the "nymphs of the West" and were often compared in mythology to the Three Graces. The apples grew on a magical tree which had golden leaves and golden bark. These apples supposedly gave eternal life to whoever ate them. Zeus, the king of the gods, had given the tree to Hera as a wedding present. She planted it in a garden at the base of Mount Atlas (which gave rise to the Atlantic Ocean, which surrounded the world near Mount Atlas). The Hesperides, daughters of Atlas, liked to pilfer from the tree, so Hera placed a ferocious serpent with 100 heads named Ladon to guard her precious tree. Hercules has three tasks to complete. First, he must capture Nereus in order to find out where the apples grow. Once Hercules captures Nereus, he must answer three riddles correctly, in **the Riddles of Nereus** problem. Hercules' second task involves wrestling the giant Antaeus. Each time Hercules throws Antaeus to the ground, the giant somehow is reenergized by the earth. We model how Hercules can defeat the giant in the **Wrestling Antaeus** problem. After that, Hercules must convince Atlas to help him obtain the apples. There are two parts to this final task (the **Hercules Has the Whole World in His Hands** problem), involving a classic circumference puzzle and determination of the earth's mass.

11.1.1. Exercise: The Riddles of Nereus

TASK: Let's suppose that the sea god Nereus forces Hercules to answer three riddles before revealing the location of the legendary tree. Here are the riddles (taken from *The Greek Anthology*):

1. "Three Hesperides were each carrying baskets of the magical apples, and in each was the same number. Nine Muses met them and asked them for apples, and they gave the same number to each Muse, and the nine Muses and three Hesperides had each of them the same number. Tell me how many they gave and how they all had the same number."

2. "Three of the Hesperides are pouring out water for the bath, sending streams into a fair-flowing tank. The one on the right, from her long-winged feet, fills it full in the sixth part of a day [1 day = 12 hours]; the one on the left, from her jar, fills it in four hours; and the one in the middle, from her bow, in just half a day. Tell me in what a short time they should fill it, pouring water from wings, bow, and jar all at once."

3. "While three Hesperides were guarding the magical tree, a thief slipped into the garden and stole some apples. On his way out he met the three Hesperides one after the other, and to each in turn he gave a half of the apples he then had, and two besides. Thus the thief managed to escape with one apple. Tell me, how many had he stolen originally?"

11.1.2. Wrestling Antaeus

TASK: Hercules must grapple with the giant Antaeus in order to pass through Libya. Each time Hercules throws Antaeus to the ground, the giant recoups energy from the earth. Hercules then lifts him off the ground, and Antaeus' strength decreases. Suppose that Hercules and Antaeus wrestle for 5 minutes before Antaeus is thrown to the ground. In the next minute, his strength is recharged. The ground provides a 1-minute pulse of energy to Antaeus; he again wrestles Hercules for 5 minutes, and his strength diminishes as before. This periodic gain and loss of energy continues throughout the battle. In the absence of the energy pulse, the rate of change in Antaeus' strength would decline at a rate proportional to three-fourths of his original strength. Model Antaeus' strength and determine if Hercules can defeat him as long as he is on land.

11.1.3. Exercise: Hercules Has the Whole World in His Hands

TASKS: This exercise is a collection of two classic problems using large numbers which were first pondered hundreds of years ago.

1. At the earth's equator, its radius is approximately 6378 kilometers. Suppose the earth is a perfect sphere and that Hercules can install

wooden poles about the equator. Then, he ties a series of ropes onto these poles, forming a circle that is concentric with the earth's equator. How much extra rope would Hercules need to add to this total rope length in order to raise the rope high enough off the ground so that Chiron the centaur could walk underneath it without touching it? Assume that Chiron stands 3 meters tall.

2. The earth's gravitational constant is 6.67×10^{-11} meters per kilogram per second squared, the acceleration due to gravity is 9.8 meters per second squared, and the radius of the earth is approximately 6378 kilometers. With only this information, determine the mass of the earth.

11.2. The Solutions

11.2.1. Exercise: The Riddles of Nereus

From Apollodorus:

When the labours had been performed in eight years and a month, Eurystheus ordered Hercules, as an eleventh labour, to fetch golden apples from the Hesperides, for he did not acknowledge the labour of the cattle of Augeas nor that of the hydra. These apples were not, as some have said, in Libya, but on Atlas among the Hyperboreans. They were presented by Earth to Zeus after his marriage with Hera, and guarded by an immortal dragon with a hundred heads, offspring of Typhon and Echidna, which spoke with many and divers sorts of voices. . . . They revealed Nereus to him, and Hercules seized him while he slept, and though the god turned himself into all kinds of shapes, the hero bound him and did not release him till he had learned from him where were the apples and the Hesperides.

TASK: Let's suppose that the sea god Nereus forces Hercules to answer three riddles before revealing the location of the legendary tree. Here are the riddles:

1. "Three Hesperides were each carrying baskets of the magical apples, and in each was the same number. Nine Muses met them and asked them for apples, and they gave the same number to each Muse, and the nine Muses and three Hesperides had each of them the same

number. Tell me how many they gave and how they all had the same number."

2. "Three of the Hesperides are pouring out water for the bath, sending streams into a fair-flowing tank. The one on the right, from her long-winged feet, fills it full in the sixth part of a day [1 day = 12 hours]; the one on the left, from her jar, fills it in four hours; and the one in the middle, from her bow, in just half a day. Tell me in what a short time they should fill it, pouring water from wings, bow, and jar all at once."

3. "While three Hesperides were guarding the magical tree, a thief slipped into the garden and stole some apples. On his way out he met the three Hesperides one after the other, and to each in turn he gave a half of the apples he then had, and two besides. Thus the thief managed to escape with one apple. Tell me, how many had he stolen originally?"

SOLUTION: The first two riddles are slight variations of arithmetical epigrams taken from Book XIV of *The Greek Anthology* (making them over 2000 years old), and the third is a rewording of an ancient Hindu problem. We will answer each in order.

1. The first riddle is just a warm-up for Hercules. If a is the number of apples each woman has at the end, then $12a = 3b$, where b is the number of apples in each basket at the beginning. So, $b = 4a$, and the 3 Hesperides each have a basket with 4 apples, which makes 12 apples in all. They each give 3 apples to the Muses, so that all 12 women have 1 apple each. Of course, any multiple of 12 apples will also answer the riddle.

2. For the second riddle, let z be the volume of water needed to fill the tank. We seek the rates needed to fill the tank, with the rate equalling the volume divided by the time (measured in hours) needed to fill the tank.

The nymph in the middle takes the most time, or half a day. Since a day is considered to be 12 hours, her rate is $\frac{z}{\frac{1}{2} \times 12} = \frac{2z}{12}$. The nymph on the left is $1\frac{1}{2}$ times faster, so her rate is $\frac{3}{2} \times \frac{2z}{12} = \frac{3z}{12}$, and the nymph on the right is 3 times faster, having a rate of $3 \times \frac{2z}{12} = \frac{6z}{12}$. The time needed is the volume divided by the sum of the three rates:

$$\text{time} = \frac{z}{\frac{2z+3z+6z}{12}} = \frac{12z}{11z} \text{ hours,}$$

so it takes $\frac{1}{11}$ of a day if all three of the Hesperides combine their rates to fill the tank.

3. Now for the third task. Perhaps this is an indication by Nereus that Hercules will be the thief. In any case, how many apples a does the thief initially steal? The first of the Hesperides receives

$$\frac{1}{2}a + 2.$$

This leaves the thief with

$$a - \left(\frac{1}{2}a + 2\right) = \frac{1}{2}a - 2$$

apples. After the encounter with the second Hesperides, the thief has

$$\frac{1}{2}\left(\frac{1}{2}a - 2\right) - 2$$

apples remaining. After the third encounter, the thief has

$$\frac{1}{2}\left(\frac{1}{2}\left(\frac{1}{2}a - 2\right) - 2\right) - 2$$

apples remaining, and we are told that this leaves the thief with only one apple. So

$$\frac{1}{2}\left(\frac{1}{2}\left(\frac{1}{2}a - 2\right) - 2\right) - 2 = 1, \qquad (11.1)$$

yielding $a = 36$ total apples stolen by the thief.

What about the general case? Define n to be the number of Hesperides. Suppose $n = 1$. Then $\frac{1}{2}a - 2 = 1$ has a solution of $a = 6$

apples. If $n = 2$, then we solve and find $a = 16$ apples. Let $n = 4$. Then we solve

$$\frac{1}{2}\left(\frac{1}{2}\left(\frac{1}{2}\left(\frac{1}{2}a - 2\right) - 2\right) - 2\right) - 2 = 1,$$

and $a = 76$. When $n = 5$, $a = 156$. We can develop a general solution. For $n = 1, 2, 3, 4, 5, \ldots$, we get $a = 6, 16, 36, 76, 156, \ldots$. Simplifying Equation 11.1 for n terms on the left-hand side, if the thief starts with

$$2^n + 2^{n+2} - 2^2$$

apples and has n encounters, he will have only one apple remaining.

Did you notice that the number of apples always ends in 6? In addition, the number of apples increases by a factor of $(2^{n-1})(10)$ as n increases from n to $n + 1$ (from 6 to 16 to 36 to 76 to 156 and so forth). Why? Rearranging terms, we see that

$$2^n + 2^{n+2} - 2^2 = 2^{n-1}(2 + 2^3) - 2^2$$
$$= 2^{n-1}(10) - 4$$
$$= 10(2^{n-1} - 1) + 6$$

is valid for $n \geq 2$. This number will always end in 6.

As a side note, in most ancient sculptures of Hercules, he is standing with three apples in his left hand. After successfully answering the three riddles, Hercules is told by Nereus where to find the Garden of the Hesperides.

11.2.2. Wrestling Antaeus

From Apollodorus:

Hercules traversed Libya. That country was then ruled by Antaeus, son of Poseidon, who used to kill strangers by forcing them to wrestle. Being forced to wrestle with him, Hercules hugged him, lifted him aloft, broke and killed him; for when he touched earth so it was that he waxed stronger, wherefore some said that he was a son of Earth.

TASK: Hercules must grapple with the giant Antaeus in order to pass through Libya. Each time Hercules throws Antaeus to the ground, the giant recoups energy from the earth. Hercules then lifts him off the ground, and Antaeus' strength decreases. Suppose that Hercules and Antaeus wrestle for 5 minutes before Antaeus is thrown to the ground. In the next minute, his strength is recharged. The ground provides a 1-minute pulse of energy to Antaeus; he again wrestles Hercules for 5 minutes, and his strength diminishes as before. This periodic gain and loss of energy continues throughout the battle. In the absence of the energy pulse, the rate of change in Antaeus' strength would decline at a rate proportional to three-fourths of his original strength. Model Antaeus' strength and determine if Hercules can defeat him as long as he is on land.

SOLUTION: We will model Antaeus' strength using an ordinary differential equation with a step "forcing" function. The easiest way to solve this is to use Laplace transforms. For a description of the background and uses of Laplace transforms, see Appendix D. To solve the differential equation we will apply the Laplace transform to the initial value problem, which hopefully will provide us with a simpler equation. Then we will solve the new algebraic equation (using partial fractions). Finally, we will apply the inverse Laplace transform to obtain the solution to the differential equation, interpreting this solution back in the context of the original problem.

How does the energy pulse affect Antaeus' strength? Let's define $a(t)$ to be his strength at time t, which is measured in minutes. Since Hercules takes strength away from Antaeus when he lifts him off the ground, the coefficient for the rate of change in Antaeus' strength (proportional to three-fourths of his original strength) is negative. This gives us the following ordinary differential equation:

$$\frac{da}{dt} = -\frac{3}{4} a(t) + f(t). \qquad (11.2)$$

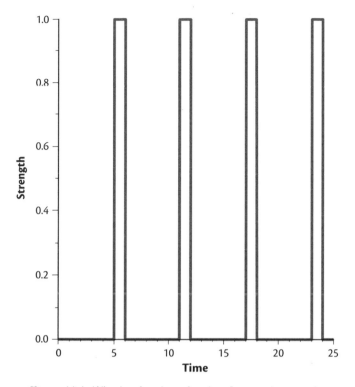

Figure 11.1. The forcing function for Antaeus' strength.

Here, $f(t)$ is the forcing function (the energy provided to Antaeus which increases his strength). Initially, we assume that Antaeus is at full strength, so $a(0) = 1$. The forcing function is a square-wave pulse. A plot of $f(t)$ is shown in Figure 11.1, and it can be modeled as

$$f(t) = \begin{cases} 0 & \text{for } 0 \le x < 5 \\ 1 & \text{for } 5 \le x < 6 \\ 0 & \text{for } 6 \le x < 11 \\ 1 & \text{for } 11 \le x < 12 \\ 0 & \text{for } 12 \le x < 17 \\ 1 & \text{for } 17 \le x < 18 \\ 0 & \text{for } 18 \le x < 23 \\ & \text{and so forth} \end{cases}$$

To solve this type of initial value problem, we use Laplace transforms. The Laplace transform of the homogeneous portion of Equation 11.2 is

$$\mathcal{L}\left[\frac{da}{dt} = -\frac{3}{4}\,a(t)\right] \Longrightarrow s\mathcal{L}[a] - a(0) = \frac{3}{4}\,\mathcal{L}[a].$$

The forcing function $f(t)$ looks like a series of unit step functions. A unit step function has a discontinuity where it jumps from 0 to 1 or from 1 back to 0. It is commonly used in differential equations to model discontinuous processes. The step function's main use is to turn other functions on and off, like a light switch. Since the function we are modeling is the addition to Antaeus' strength, which is equal to either 1 or 0, our model needs to turn the value of the strength on and off at the appropriate time. For example, $f(t) \cdot \text{step}(t, \alpha)$ turns the function $f(t)$ on at $t = \alpha$, setting the function to a value of 1. Further, $f(t) \cdot \{\text{step}(t, \alpha) - \text{step}(t, \beta)\}$ turns $f(t)$ on at $t = \alpha$ and off at $t = \beta$. Our function $f(t)$ can therefore be modeled as

$$f(t) = 1 \times \{\text{step}(t, 5) - \text{step}(t, 6) + \text{step}(t, 11)$$
$$-\text{step}(t, 12) + \text{step}(t, 17) - \text{step}(t, 18) + \cdots\}.$$

The Laplace transform of $\text{step}(t, 5)$ as a function of s is

$$\mathcal{L}\left[\text{step}(t, 5)\right] = \frac{1}{s} \times e^{-5s},$$

and when we add up all the terms, the Laplace transform of $f(t)$ becomes

$$\mathcal{L}[f(t)] = \frac{1}{s}\left[e^{-5s} - e^{-6s} + e^{-11s} - e^{-12s} + e^{-17s} - e^{-18s} + \cdots\right].$$

Instead of using this continuing sum, let's truncate the forcing function at $t = 18$ and see if we can determine any trend. Thus, the transformed differential equation becomes

$$s\mathcal{L}[a] - a(0) = -\frac{3}{4}\,\mathcal{L}[a] + \frac{1}{s}\left[e^{-5s} - e^{-6s}\right.$$
$$\left. +e^{-11s} - e^{-12s} + e^{-17s} - e^{-18s} + \cdots\right]. \qquad (11.3)$$

Assign the value of $a(0) = 1$ and solve Equation 11.3 for $\mathcal{L}[a]$.

$$\left(s + \frac{3}{4}\right)\mathcal{L}[a] = 1 + \frac{1}{s}\left[e^{-5s} - e^{-6s} + e^{-11s} - e^{-12s}\right.$$

$$\left. + e^{-17s} - e^{-18s} + \cdots\right].$$

So,

$$\mathcal{L}[a] = \frac{1}{s + \frac{3}{4}} + \frac{1}{s\left(s + \frac{3}{4}\right)}\left[e^{-5s} - e^{-6s} + e^{-11s} - e^{-12s}\right.$$

$$\left. + e^{-17s} - e^{-18s} + \cdots\right]. \tag{11.4}$$

All that's left to do now is to take the inverse Laplace transform of the right-hand side of Equation 11.4. The inverse transforms take the following forms:

$$\mathcal{L}^{-1}\left[\frac{1}{s - k}\right] = e^{kt},$$

$$\mathcal{L}^{-1}\left[\frac{1}{s(s - k)}e^{-rs}\right] = \frac{1}{k}\,\text{step}(t, r)\left(1 - e^{k(t-r)}\right).$$

So, the solution to the initial value problem transforms to

$$a(t) = e^{-\frac{3}{4}t} + \frac{4}{3}\left[\text{step}(t, 5)\left(1 - e^{-0.75t+3.75}\right)\right]$$

$$- \frac{4}{3}\left[\text{step}(t, 6)\left(1 - e^{-0.75t+4.50}\right)\right]$$

$$+ \frac{4}{3}\left[\text{step}(t, 11)\left(1 - e^{-0.75t+8.25}\right)\right]$$

$$- \frac{4}{3}\left[\text{step}(t, 12)\left(1 - e^{-0.75t+9.00}\right)\right]$$

$$+ \frac{4}{3}\left[\text{step}(t, 17)\left(1 - e^{-0.75t+12.75}\right)\right]$$

$$- \frac{4}{3}\left[\text{step}(t, 18)\left(1 - e^{-0.75t+13.50}\right)\right]$$

$$+ \cdots . \tag{11.5}$$

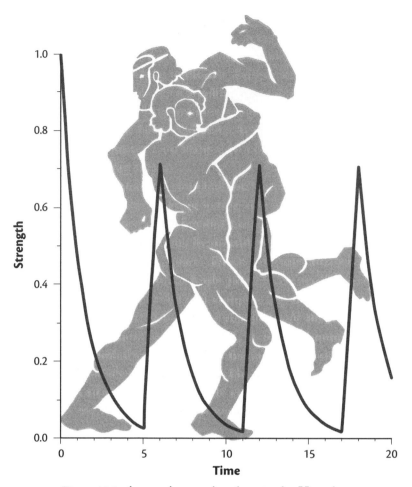

Figure 11.2. Antaeus' strength as he wrestles Hercules.

Equation 11.5 represents the changing strength of the giant Antaeus as the wrestling match progresses, and this solution is plotted in Figure 11.2. You can see that there is exponential decay while Hercules holds Antaeus above the ground for the first 5 minutes. Then, as his strength is at almost nothing, Antaeus is thrown to the ground, and he regains his strength (notice the steep rate of change). Hercules wrestles again with Antaeus, and the cycle repeats and repeats. The only way Hercules will win is if he can either strangle Antaeus while he is in the air or throw him into the water, where he will not gain strength from the earth.

11.2.3. Exercise: Hercules Has the Whole World in His Hands

From Apollodorus:

Now Prometheus had told Hercules not to go himself after the apples but to send Atlas, first relieving him of the burden of the sphere; so when he was come to Atlas in the land of the Hyperboreans, he took the advice and relieved Atlas. But when Atlas had received three apples from the Hesperides, he came to Hercules, and not wishing to support the sphere he said that he would himself carry the apples to Eurystheus, and bade Hercules hold up the sky in his stead. Hercules promised to do so, but succeeded by craft in putting it on Atlas instead. For at the advice of Prometheus he begged Atlas to hold up the sky till he should put a pad on his head. When Atlas heard that, he laid the apples down on the ground and took the sphere from Hercules. And so Hercules picked up the apples and departed.

TASKS: This exercise is a collection of two classic problems using large numbers which were first pondered hundreds of years ago.

1. At the earth's equator, its radius is approximately 6378 kilometers. Suppose the earth is a perfect sphere and that Hercules can install wooden poles about the equator. Then, he ties a series of ropes onto these poles, forming a circle that is concentric with the earth's equator. How much extra rope would Hercules need to add to this total rope length, in order to raise the rope high enough off the ground so that Chiron the centaur could walk underneath it without touching it? Assume that Chiron stands 3 meters tall.

2. The earth's gravitational constant is 6.67×10^{-11} meters per kilogram per second squared, the acceleration due to gravity is 9.8 meters per second squared, and the radius of the earth is approximately 6378 kilometers. With only this information, determine the mass of the earth.

SOLUTION:

1. The circumference of the earth is $C = 2\pi r = 2\pi \times 6378 \approx$ 40,074.156 kilometers. By adding 3 meters (0.003 kilometer) to the radius, the circumference becomes

$$C = 2\pi \times 6378.003 = 40,074.175 \text{ kilometers.}$$

By adding 3 meters to the circle's radius, Hercules needs to add $2\pi \times 3 = 6\pi$ meters of rope to the current total length. Really. Less than 20 meters of rope will work.

Jesper Parnevik, a professional golfer, pondered this problem at the 1999 Standard Life Golf Tournament, saying, "If you put a rope around the earth and measure it at 26,000 miles and then put another rope three feet above the surface, how much longer is that rope? I missed a two-foot putt thinking about it and suddenly had a four-footer for bogey."

2. To determine the mass of the earth, all we need are Newton's Law of Gravity and Newton's Second Law of Motion. The Law of Gravity works for any two objects with mass. If two objects, say A and E (for example, the earth), with masses m_A and m_E, respectively, are situated a distance r apart, then the force directed toward each object is

$$F = \frac{G\, m_A m_E}{r^2}, \qquad (11.6)$$

where G is the universal gravitational constant. Since it is directed toward each object, this force is always attractive. Newton stated that his gravity law works for any two objects with mass. For example, it applies for any motion on the earth, as well as any motion in the universe. Newton's Second Law of Motion relates an object's mass, acceleration, and force. Again, if we have an object A, the gravitational force acting on it can be given by

$$F = m_A\, g, \qquad (11.7)$$

where m_A is the mass of the body and g is the acceleration due to gravity. This law allows us to measure how velocities change when forces are applied since acceleration is the rate of change of velocity with respect to time. We can now equate Equations 11.6 and 11.7:

$$\frac{G m_A m_E}{r^2} = m_A g,$$

$$\frac{G m_E}{r^2} = g,$$

$$m_E = \frac{g r^2}{G}.$$

Now we can calculate the mass of the earth. The value of G was calculated by Henry Cavendish in 1798, and it was determined to be 6.67×10^{-11} meter^3/(kg sec^2), and $a = 9.8$ meters/sec^2. Converting the radius of the earth to meters, we can now calculate the mass of the earth:

$$m_E = \frac{9.8 \times (6,378,137)^2}{6.67 \times 10^{-11}} \approx 6.0 \times 10^{24} \text{ kilograms.}$$

How much work does Hercules do in placing the earth on his massive shoulders? Apollodorus is somewhat vague about how Hercules lifts or holds the earth, so we will be, too. The physics involved in moving an object the size of the earth even a few meters requires a Herculean effort. No wonder Atlas was tired of holding up the earth.

The Twelfth Labor: Cerberus

From Apollodorus:

A twelfth labour imposed on Hercules was to bring Cerberus from Hades. Now this Cerberus had three heads of dogs, the tail of a dragon, and on his back the heads of all sorts of snakes. When Hercules was about to depart to fetch him, he went to Eumolpus at Eleusis, wishing to be initiated. However it was not then lawful for foreigners to be initiated: since he proposed to be initiated as the adoptive son of Pylius. But not being able to see the mysteries because he had not been cleansed of the slaughter of the centaurs, he was cleansed by Eumolpus and then initiated. And having come to Taenarum in Laconia, where is the mouth of the descent to Hades, he descended through it. But when the souls saw him, they fled, save Meleager and the Gorgon Medusa. And Hercules drew his sword against the Gorgon, as if she were alive, but he learned from Hermes that she was an empty phantom. And being come near to the gates of Hades he found Theseus and Pirithous, him who wooed Persephone in wedlock and was therefore bound fast. And when they beheld Hercules, they stretched out their hands as if they should be raised from the dead by his might. And Theseus, indeed, he took by the hand and raised up, but when he would have brought up Pirithous, the earth quaked and he let go. And he rolled away also the stone of Ascalaphus. And wishing to provide the souls with blood, he slaughtered one of the kine of Hades. But Menoetes, son of Ceuthonymus, who tended the king, challenged Hercules to wrestle, and, being seized round the middle, had his ribs broken; howbeit, he was let off at the request of Persephone. When Hercules asked Pluto for Cerberus, Pluto ordered him to take the animal provided he mastered him without the use of the weapons which he carried. Hercules found him at the gates of Acheron, and, cased in his cuirass and

134

covered by the lion's skin, he flung his arms round the head of the brute, and though the dragon in its tail bit him, he never relaxed his grip and pressure till it yielded. So he carried it off and ascended through Troezen. But Demeter turned Ascalaphus into a short-eared owl, and Hercules, after showing Cerberus to Eurystheus, carried him back to Hades.

12.1. The Tasks

Hercules has two tasks to complete. He finds his way to the entrance of Hades. Once at Taenarum, Hercules must determine the direction that will allow him to descend as quickly as possible (**The Descent into the Underworld** problem). As he goes deeper and deeper into Hades, he passes several souls who stretch out their hands, asking him to save them. However, Hercules is intent on completing his last labor. Pluto insists that Hercules not use any weapons when capturing Cerberus. Our hero must wrestle and strangle the beast with his bare hands (**The Fight with Cerberus** problem), in effect cutting off the oxygen supply to all three of the dog's heads. Once Hercules subdues the brute, he can carry it off to Eurystheus.

12.1.1. The Descent into the Underworld

TASK: Hercules arrives at Taenarum, "the mouth of the descent to Hades," from the west (the negative x-direction). The terrain can be characterized by the following function: $f(x, y) = -2x + y^2 - 2xy$, measured in meters, with the entrance at Taenarum at the point $(2, 1)$. If Hercules continues from the entrance in a northeasterly direction, say from the entrance to the point $(4, 3)$, what is his rate of change at the entrance and is he initially climbing up or down? In which direction should Hercules proceed to climb down as fast as possible? Now, suppose Hercules tries to enter the Underworld at a different point, say $(2, -1)$. Describe his path (in which direction must he travel to descend?).

12.1.2. The Fight with Cerberus

TASK: Hercules is allowed to capture the beast by using only his bare hands, so he chooses to wrestle the animal and strangle each of its heads. A typical blood flow rate through the carotid artery to the brain is about 6 milliliters (ml) per second. Cerberus has three brains, so Hercules must attempt to reduce the flow in each head. Assume that by strangling one head at a time, Hercules can reduce the blood flow by 7 percent of the current amount of blood in the brain per second. Once the amount of blood in one head falls below 2 ml, he grabs a second head and begins strangling it. The head he lets go of then shows an increase in blood flow at the rate of 0.05 ml per second. When will the amount of blood in all three brains be less than 2.5 ml, causing Cerberus to become passive and yield to Hercules?

12.2. The Solutions

12.2.1. The Descent into the Underworld

From Apollodorus:

And having come to Taenarum in Laconia, where is the mouth of the descent to Hades, [Hercules] descended through it. But when the souls saw him, they fled, save Meleager and the Gorgon Medusa. And Hercules drew his sword against the Gorgon, as if she were alive, but he learned from Hermes that she was an empty phantom. And being come near to the gates of Hades he found Theseus and Pirithous, him who wooed Persephone in wedlock and was therefore bound fast. And when they beheld Hercules, they stretched out their hands as if they should be raised from the dead by his might. And Theseus, indeed, he took by the hand and raised up, but when he would have brought up Pirithous, the earth quaked and he let go. And he rolled away also the stone of Ascalaphus.

TASK: Hercules arrives at Taenarum, "the mouth of the descent to Hades," from the west (the negative x-direction). The terrain can be characterized by the following function: $f(x, y) = -2x + y^2 - 2xy$, measured in meters, with the entrance at Taenarum at the point $(2, 1)$. If Hercules continues from the entrance in a northeasterly direction, say from the entrance to the point $(4, 3)$, what is his rate of change at the entrance and is he initially climbing up or down? In which direction should Hercules proceed to climb down as fast as possible?

Now, suppose Hercules tries to enter the Underworld at a different point, say $(2, -1)$. Describe his path (in which direction must he travel to descend?).

SOLUTION: The Underworld was the kingdom of Hades and his wife, Persephone. Depending on how a person lived his or her life, he or she might or might not experience never-ending punishment in Hades. All souls, whether good or bad, were destined for the kingdom of Hades.

We are given that the terrain at Taenarum can be characterized by the function $f(x, y) = -2x + y^2 - 2xy$, with the entrance at a point $P = (2, 1)$. To determine the path of steepest descent, we need to find the gradient of $f(x, y)$, which is given by

$$\nabla f(x, y) = \frac{\partial f}{\partial x}\mathbf{i} + \frac{\partial f}{\partial y}\mathbf{j}$$
$$= (-2 - 2y)\mathbf{i} + (-2x + 2y)\mathbf{j}. \qquad (12.1)$$

Equation 12.1 gives the gradient at any point on the surface $f(x, y)$. The gradient is the rate of change of a function of multiple variables, and the gradient vector gives the direction of fastest increase of the function. A gradient vector field plot superimposed on the contour plot of the function is shown in Figure 12.1.

The vectors point uphill in the direction of steepest ascent. Notice that the gradient vectors are perpendicular to the level curves of the surface $f(x, y)$. Now we check the gradient at the initial point $(2, 1)$, where $\nabla f(x, y) = -4\mathbf{i} - 2\mathbf{j}$. We can clearly see that Hercules will be climbing down, but at what rate? In the direction of $\mathbf{PQ}$, we need to calculate the directional derivative of $f(x, y)$ at the point $(2, 1)$, where Q is the point $(4, 3)$. We see that

$$\mathbf{PQ} = 2\mathbf{i} + 2\mathbf{j}.$$

A unit vector in the direction of $\mathbf{PQ}$ is given by

$$\mathbf{u} = \frac{\mathbf{PQ}}{|\mathbf{PQ}|} = \frac{2\mathbf{i} + 2\mathbf{j}}{\sqrt{2^2 + 2^2}} = \frac{\sqrt{2}}{2}\mathbf{i} + \frac{\sqrt{2}}{2}\mathbf{j}.$$

Now that we have this unit vector, the rate of change of $f(x, y)$ in the direction from the entrance at P to the point Q is determined by the

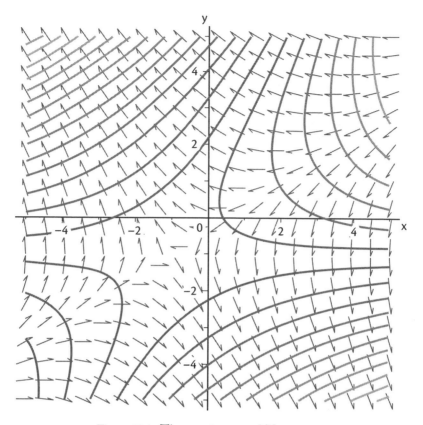

Figure 12.1. The terrain around Taenarum.

directional derivative:

$$D_u f(2,1) = \nabla f(2,1) \cdot \mathbf{u}$$

$$= (-4\mathbf{i} - 2\mathbf{j}) \cdot \left(\frac{\sqrt{2}}{2}\mathbf{i} + \frac{\sqrt{2}}{2}\mathbf{j} \right)$$

$$= -2\sqrt{2} - \sqrt{2}$$

$$= -3\sqrt{2}.$$

Physically, this is the slope of the tangent line to the curve of intersection of our surface (the terrain represented by $f(x, y)$) with a vertical plane through P in the direction of $\mathbf{u}$.

In which direction should Hercules proceed to climb down as fast as possible? The greatest rate of change is equal to the magnitude of the gradient, or $|\nabla f(x, y)|$. This is given by

$$|\nabla f(x, y)| = \sqrt{(-4)^2 + (-2)^2} = 4.472,$$

which indicates that the rate of descent is 4.472. In order to descend as rapidly as possible, Hercules should follow the path designated by the gradient. He should stay perpendicular to all the contour lines (level curves). From Figure 12.1, that would mean heading in a northeasterly direction, not quite as north as to the point Q but following the arrows in a reverse direction.

Suppose Hercules tries to enter the Underworld at the point $(x, y) = (2, -1)$. Let's discuss this qualitatively. Look at Figure 12.1 again. At the point $(2, -1)$, the vector is pointing in the negative y-direction, indicating that the negative y-direction is uphill. In order to *descend* into the Underworld, Hercules must head in a due north direction and then start to turn northeast, as outlined in the first part of this task. Once he gets past all the souls of Hades, Hercules asks Pluto for Cerberus, and Pluto orders him to take the animal provided he captures the great beast without the use of any weapons.

12.2.2. The Fight with Cerberus

From Apollodorus:

When Hercules asked Pluto for Cerberus, Pluto ordered him to take the animal provided he mastered him without the use of the weapons which he carried. Hercules found him at the gates of Acheron, and, cased in his cuirass and covered by the lion's skin, he flung his arms round the head of the brute, and though the dragon in its tail bit him, he never relaxed his grip and pressure till it yielded. So he carried it off and ascended through Troezen.

TASK: Hercules is allowed to capture the beast by using only his bare hands, so he chooses to wrestle the animal and strangle each of its heads. A typical blood flow rate through the carotid artery to the brain

is about 6 milliliters (ml) per second. Cerberus has three brains, so
Hercules must attempt to reduce the flow in each head. Assume that
by strangling one head at a time, Hercules can reduce the blood flow by
7 percent of the current amount of blood in the brain per second. Once
the amount of blood in one head falls below 2 ml, he grabs a second
head and begins strangling it. The head he lets go of then shows an
increase in blood flow at the rate of 0.05 ml per second. When will the
amount of blood in all three brains be less than 2.5 ml, causing Cerberus
to become passive and yield to Hercules?

SOLUTION: How does one defeat a three-headed dog? Harry Potter
had to get past Fluffy, Hagrid's three-headed dog, in *Harry
Potter and the Sorceror's Stone*. During a conversation with Harry,
Hagrid says, "How many three-headed dogs d'yeh meet, even around
Hogwarts? ... Fluffy's a piece o' cake if yeh know how to calm him
down, jus' play him a bit o' music an' he'll go straight off ter sleep—I
shouldn'ta told yeh that!" Hercules knows that Cerberus will not be so
easily vanquished.

Cerberus guarded the gate to Hades and ensured that the spirits
of the dead could enter but that none could exit (additionally, no
living person was to come into Hades). From the spittle of the dog
which fell upon earth, the first poisonous plants had supposedly been
born, including deadly aconite. In Greek mythology, Orpheus, one of
the more famous poets and musicians and the inventor of the lyre,
lulled Cerberus to sleep with his lyre, similar to the taming of Fluffy.
However, Hercules takes the brute force approach and attempts to
strangle the three-headed beast. In doing so, he decreases Cerberus'
blood supply to the three brains, causing ischemia. Ischemia is the
medical term for insufficiency of blood supply to the brain. The lack
of sufficient blood flow deprives the brain of valuable oxygen but also
prevents the removal of harmful lactic acid and other toxins from the
brain. If the ischemia becomes prolonged and severe, cellular elements
may die in the brain, creating the possibly fatal condition known as an
infarction.

We are told that Hercules dons his protective lion's skin and grabs
one head of Cerberus, attempting to strangle it, thereby reducing blood
flow to that head's brain. Initially, the amount of blood in the brain

time	Head 1	Head 2	Head 3
0	6.0000	6.0000	6.0000
1	5.5944	6.0000	6.0000
2	5.2161	6.0000	6.0000
3	4.8635	6.0000	6.0000
4	4.5347	6.0000	6.0000
5	4.2281	6.0000	6.0000
6	3.9423	6.0000	6.0000
7	3.6758	6.0000	6.0000
8	3.4273	6.0000	6.0000
9	3.1956	6.0000	6.0000
10	2.9795	6.0000	6.0000
11	2.7781	6.0000	6.0000
12	2.5903	6.0000	6.0000
13	2.4151	6.0000	6.0000
14	2.2519	6.0000	6.0000
15	2.0996	6.0000	6.0000
16	1.9577	6.0000	6.0000
17	2.0077	5.5944	6.0000
18	2.0577	5.2161	6.0000
19	2.1077	4.8635	6.0000
20	2.1577	4.5347	6.0000
21	2.2077	4.2281	6.0000
22	2.2577	3.9423	6.0000
23	2.3077	3.6758	6.0000
24	2.3577	3.4273	6.0000
25	2.4077	3.1956	6.0000
26	2.4577	2.9795	6.0000
27	2.5077	2.7781	6.0000
28	2.5577	2.5903	6.0000
29	2.6077	2.4151	6.0000
30	2.6577	2.2519	6.0000
31	2.7077	2.0996	6.0000
32	2.7577	1.9577	6.0000
33	2.8077	2.0077	5.5944
34	2.8577	2.0577	5.2161

time	Head 1	Head 2	Head 3
35	2.9077	2.1077	4.8635
36	2.9577	2.1577	4.5347
37	3.0077	2.2077	4.2281
38	3.0577	2.2577	3.9423
39	3.1077	2.3077	3.6758
40	3.1577	2.3577	3.4273
41	3.2077	2.4077	3.1956
42	3.2577	2.4577	2.9795
43	3.3077	2.5077	2.7781
44	3.3577	2.5577	2.5903
45	3.4077	2.6077	2.4151
46	3.4577	2.6577	2.2519
47	3.5077	2.7077	2.0996
48	3.5577	2.7577	1.9577
49	3.3172	2.8077	2.0077
50	3.0929	2.8577	2.0577
51	2.8838	2.9077	2.1077
52	2.6888	2.9577	2.1577
53	2.5071	3.0077	2.2077
54	2.3376	3.0577	2.2577
55	2.1795	3.1077	2.3077
56	2.0322	3.1577	2.3577
57	1.8948	3.2077	2.4077
58	1.9448	2.9908	2.4577
59	1.9948	2.7886	2.5077
60	2.0448	2.6001	2.5577
61	2.0948	2.4243	2.6077
62	2.1448	2.2604	2.6577
63	2.1948	2.1076	2.7077
64	2.2448	2.1576	2.5246
65	2.2948	2.2076	2.3539
66	2.3448	2.2576	2.1948
67	2.3948	2.3076	2.0464
68	2.4448	2.3576	1.9081

Figure 12.2. Spreadsheet of the blood flow.

flows at 6 ml per second (a typical blood flow rate through the carotid artery is 350 ml per minute) Let's assume that there is 6 ml of blood in the brain when Hercules begins to grab one head. During strangulation, Hercules reduces that amount by 7 percent per second until the blood in one brain is less than 2 ml. We define $b(t)$ to be the amount of blood in the brain (measured in milliliters) at time t (measured in seconds). This situation is described by the differential equation

$$\frac{db}{dt} = -0.07b. \tag{12.2}$$

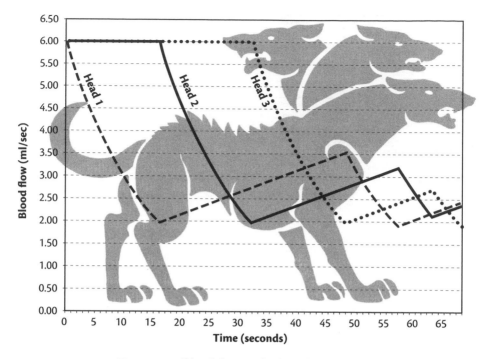

Figure 12.3. Blood flow in Cerberus' three heads.

The general solution is $b(t) = ce^{-0.07t}$, where c is an unknown constant. Applying the initial condition of $b(0) = 6$ yields a particular solution of $b(t) = 6e^{-0.07t}$. After 16 seconds, b(16) = 1.9577 ml (see Figure 12.2).

After 16 seconds, Hercules releases head 1 and begins to strangle head 2. It takes another 16 seconds for its amount of blood to fall below 2 ml. However, during this time (32 seconds), the blood flow in head 1 has been increasing at a rate of 0.05 ml per second, so it has climbed to 2.7577 ml. The blood flow in head 3 is still at 6 ml. Hercules grabs head 3, and we see that after 48 seconds, the blood flow in this head is below 2 ml. The blood flow in head 2 has risen to 2.7577 ml, and the blood flow in head 1 has risen to 3.5577 ml. So, Hercules releases the third head and grabs head 1 again, causing a 7 percent decrease in blood flow. It takes 9 seconds for this amount to again fall below 2 ml, but the other two heads are seeing increases in amounts of blood. Hercules begins strangling head 2 at 58 seconds and then head 3 at

1			3					7
	7	5					7	
	7	6			2	3		
5			7			6		
				7				
		2			7			1
		4	1			5	7	
	7					4	7	
7					6			3

Figure 12.4. Cerberus sudoku.

64 seconds. We see that after 65 total seconds, just over a minute, the amounts of blood in all three heads of Cerberus are below the 2.5-ml level. Cerberus yields to Hercules, who carries the beast to Eurystheus. The blood flow to all three heads is shown graphically in Figure 12.3.

Is this realistic? Hercules does not want to kill Cerberus. It does not take long to substantially decrease the blood flow to the brain by strangulation since closure of blood vessels often leads to a stroke or other cerebral damage. Hercules merely applies enough pressure to cause the dog to lose some consciousness and yield to his strength. For more information on the brain, a number of great web sites exist, including http://faculty.washington.edu/chudler/facts.html.

12.3. The Cerberus Sudoku Puzzle

With the completion of the Twelfth Labor, Hercules has cleansed his soul. He deserves to celebrate. You deserve to celebrate! Enjoy the following sudoku puzzle (see Figure 12.4). This is a "three-head" sudoku puzzle. The rules are the same is in normal sudoku but with each block, row, and column containing the numbers 1 through 6 exactly once and the number 7 three times (one for each of Cerberus' heads). Good luck!

APPENDIXES

The Labors and Subject Areas of Mathematics

A.1. Subject Areas by Labors and Tasks

The following table lists the labors and tasks with their associated subject areas of mathematics.

Labor and Tasks	Area of Mathematics
First Labor: The Nemean Lion	
Shooting an Arrow	Integral calculus
Hercules Closes the Cave Mouth	Geometry, algebra
Exercise: Zeus Makes a Deal	Probability
Second Labor: The Lernean Hydra	
One Head Replaced by Two	Difference equations
Cauterizing the Hydra	Probability
Third Labor: The Hind of Ceryneia	
Optimizing the Hind's Journey	Differential calculus
Cerynitian Work	Integral calculus
Exercise: Work with a Variable Force	Integral calculus
Fourth Labor: The Erymanthian Boar	
Exercise: The Centaur's Wine	Algebra
Chiron's Poison	Difference equations
The Capture of the Boar	Differential equations
Fifth Labor: The Augean Stables	
The Herds of Augeas	Algebra
Exercise: Hydrostatic Pressure on the Stable Walls	Integral calculus
Cleaning the Stables with Torricelli	Differential equations

Sixth Labor: The Stymphalian Birds
The Spiral of Archimedes   Integral calculus
Resonating Castanets   Differential equations
Exercise: Monte Carlo Shooting Scheme   Simulations

Seventh Labor: The Cretan Bull
Exercise: Riding the Bull   Statistics
The Marathon Attacks   Probability,
                                                integral calculus

Eighth Labor: The Horses of Diomedes
Driving the Mares to the Sea   Differential calculus
Hercules' Slingshot   Multivariable calculus
Exercise: The City of Abdera   Differential equations

Ninth Labor: The Belt of Hippolyte
The Sons of Minos versus Hercules   Algebra, combinatorics
The Amazons and the Spread of a Rumor   Difference equations
Exercise: Hercules and the Kraken   Differential calculus

Tenth Labor: Geryon's Cattle
The Pillars of Hercules   Differential calculus
The Golden Goblet   Geometry, trigonometry
Hera Sends the Gadflies   Differential equations
Blocking the River Strymon   Geometry

**Eleventh Labor: The Apples
of the Hesperides**
Exercise: The Riddles of Nereus   Algebra
Wrestling Antaeus   Differential equations
                                                (Laplace transforms)
Exercise: Hercules Has the Whole   Algebra
   World in His Hands

Twelfth Labor: Cerberus
The Descent into the Underworld   Multivariable calculus
The Fight with Cerberus   Differential equations

A.2. Tasks by Subject Area

This table lists the subject areas of mathematics in alphabetical order along with the tasks and the labor dealing with each of these areas.

Area of Mathematics	Task Name (Labor)
Algebra	Hercules Closes the Cave Mouth (1)
	Exercise: The Centaurs' Wine (4)
	The Herds of Augeas (5)
	The Sons of Minos versus Hercules (9)
	Exercise: The Riddles of Nereus (11)
	Exercise: Hercules Has the Whole World In His Hands (11)
Combinatorics	The Sons of Minos versus Hercules (9)
Difference equations	One Head Replaced by Two (2)
	Chiron's Poison (4)
	The Amazons and the Spread of a Rumor (9)
Differential calculus	Optimizing the Hind's Journey (3)
	Driving the Mares to the Sea (8)
	Exercise: Hercules and the Kraken (9)
Differential equations	Chiron's Poison (4)
	The Capture of the Boar (4)
	Cleaning the Stables With Torricelli (5)
	Resonating Castanets (6)
	Exercise: The City of Abdera (8)
	Hera Sends the Gadflies (10)
	Wrestling Antaeus (11)
	The Fight with Cerberus (7)

Geometry	Hercules Closes the Cave Mouth (1)
	The Golden Goblet (10)
	Blocking the River Strymon (10)
Integral calculus	Shooting an Arrow (1)
	Cerynitian Work (3)
	Exercise: Work with a Variable Force (3)
	Exercise: Hydrostatic Pressure
	on the Stable Walls (5)
	The Spiral of Archmides (6)
	The Marathon Attacks (7)
Multivariable calculus	Hercules' Slingshot (8)
	The Descent into the Underworld (12)
Probability	Exercise: Zeus Makes a Deal (1)
	Cauterizing the Hydra (2)
	The Marathon Attacks (7)
Simulations	Exercise: Monte Carlo Shooting Scheme (6)
Statistics	Exercise: Riding the Bull (7)
Trigonometry	The Golden Goblet (10)

B

Hercules before the Labors

The story surrounding Hercules cannot be told starting from the middle. Therefore, I wanted to share an account of Hercules before he began his labors, from his birth until the time he murdered his family and was told to serve Eurystheus for 12 years. It is a short but important part of the story, and Apollodorus tells it in Book Two, Sections 4.8 through 4.12, which is in Volume I of *The Library*.

B.1. Hercules' Background

From Apollodorus:

But before Amphitryon reached Thebes, Zeus came by night and prolonging the one night threefold he assumed the likeness of Amphitryon and bedded with Alcmena and related what had happened concerning the Teleboans. But when Amphitryon arrived and saw that he was not welcomed by his wife, he inquired the cause; and when she told him that he had come the night before and slept with her, he learned from Tiresias how Zeus had enjoyed her. And Alcmena bore two sons, to wit, Hercules, whom she had by Zeus and who was the elder by one night, and Iphicles, whom she had by Amphitryon. When the child was eight months old, Hera desired the destruction of the babe and sent two huge serpents to the bed. Alcmena called Amphitryon to her help, but Hercules arose and killed the serpents by strangling them with both his hands. However, Pherecydes says that it was Amphitryon who put the serpents in the bed, because he would know which of the two children was his, and that when Iphicles fled, and Hercules stood his ground, he knew that Iphicles was begotten of his body.

Hercules was taught to drive a chariot by Amphitryon, to wrestle by Autolycus, to shoot with the bow by Eurytus, to fence by Castor, and to play the lyre by Linus. This Linus was a brother of Orpheus; he came to Thebes and became a Theban, but was killed by Hercules with a blow of the lyre; for being struck by him, Hercules flew into a rage and slew him. When he was tried for murder, Hercules quoted a law of Rhadamanthys, who laid it down that whoever defends himself against a wrongful aggressor shall go free, and so he was acquitted. But fearing he might do the like again, Amphitryon sent him to the cattle farm; and there he was nurtured and outdid all in stature and strength. Even by the look of him it was plain that he was a son of Zeus; for his body measured four cubits, and he flashed a gleam of fire from his eyes; and he did not miss, neither with the bow nor with the javelin. While he was with the herds and had reached his eighteenth year he slew the lion of Cithaeron, for that animal, sallying from Cithaeron, harried the kine of Amphitryon and of Thespius.

Now this Thespius was king of Thespiae, and Hercules went to him when he wished to catch the lion. The king entertained him for fifty days, and each night, as Hercules went forth to the hunt, Thespius bedded one of his daughters with him (fifty daughters having been borne to him by Megamede, daughter of Arneus); for he was anxious that all of them should have children by Hercules. Thus Hercules, though he thought that his bed-fellow was always the same, had intercourse with them all. And having vanquished the lion, he dressed himself in the skin and wore the scalp as a helmet.

As he was returning from the hunt, there met him heralds sent by Erginus to receive the tribute from the Thebans. Now the Thebans paid tribute to Erginus for the following reason. Clymenus, king of the Minyans, was wounded with a cast of a stone by a charioteer of Menoeceus, named Perieres, in a precinct of Poseidon at Onchestus; and being carried dying to Orchomenus, he with his last breath charged his son Erginus to avenge his death. So Erginus marched against Thebes, and after slaughtering not a few of the Thebans he concluded a treaty with them, confirmed by oaths, that they should send him tribute for twenty years, a hundred kine every year. Falling in with the heralds on their way to Thebes to demand this tribute, Hercules outraged them; for he cut off their ears and noses and hands, and having fastened them by ropes from their necks, he told them to carry that tribute to Erginus and the Minyans. Indignant at this outrage, Erginus marched against Thebes. But Hercules, having received weapons from Athena and taken the command, killed Erginus, put the Minyans to flight, and compelled them to pay double the tribute to the Thebans. And it chanced that in the fight Amphitryon fell fighting bravely. And Hercules received from Creon his eldest daughter Megara as a prize of valor, and by her he had three sons, Therimachus, Creontiades, and Deicoon. But Creon gave his younger daughter to Iphicles, who already had a son Iolaus by Automedusa, daughter of Alcathus. And Rhadamanthys, son of Zeus, married Alcmena after the death of Amphitryon, and dwelt as an exile at Ocaleae in Boeotia. Having first learned from Eurytus the art of archery, Hercules received a sword from Hermes, a bow and arrows from Apollo, a golden breastplate from Hephaestus, and a robe from Athena; for he had himself cut a club at Nemea.

Now it came to pass that after the battle with the Minyans Hercules was driven mad through the jealousy of Hera and flung his own children, whom he had by Megara, and two children of Iphicles into the fire; wherefore he condemned himself to exile, and was purified by Thespius, and repairing to Delphi he inquired of the god where he should dwell. The Pythian priestess then first called him Hercules, for hitherto he was called Alcides. And she told him to dwell in Tiryns, serving Eurystheus for twelve years and to perform the ten labours imposed on him, and so, she said, when the tasks were accomplished, he would be immortal.

The Authors of the Hercules Myth

C.1. The Authors

There are at least four main authors from classical times who wrote about Hercules and his labors: Apollodorus, Diodorus Siculus, Hyginus, and Euripides. The epic poet Homer also mentions Hercules in *The Iliad*. In addition, there are several passages in *The Greek Anthology* that refer to Hercules and his 12 labors. Book XVI of *The Greek Anthology* contains epigrams by several different authors. Each epigram is a short verse describing the wondrous deeds of Hercules. Number 92 has an anonymous author and is entitled "The Labours of Heracles":

> First, in Nemea he slew the might lion. Secondly, in Lerna he destroyed the many-necked hydra. Thirdly, after this he killed the Erymanthean boar. Next, in the fourth place, he captured the hind with the golden horns. Fifthly, he chased away the Stymphalian birds. Sixthly, he won the Amazon's bright girdle. Seventhly, he cleaned out the abundant dung of Augeas. Eighthly, he drove away from Crete the fire-breathing bull. Ninthly, he carried off from Thrace the horses of Diomede. Tenthly, he brought from Erythea the oxen of Geryon. Eleventhly, he led up from Hades the dog Cerberus. Twelfthly, he brought to Greece the golden apples. In the thirteenth place he had this terrible labour: in one night he lay with fifty maidens.

This epigram contains the same 12 labors as the writings of Apollodorus, but in a slightly different order. It also adds a story about Thespius, who was king of Thespiae. Hercules went to him when he wished to catch the lion. The king entertained Hercules for 50 days, and each night, as Hercules went forth to the hunt, Thespius sent one of his 50 daughters with him, for he was anxious that all of them should have children by Hercules. After Hercules vanquished the lion, he dressed himself in the skin and wore the scalp as a helmet. This is valuable information, as almost all depictions of Hercules show him wearing the skin of the lion.

Number 93 (by Philippus, an author from the second century) is another epigram that enumerates the labors:

> The Nemean monster, and the Hydra dire
> I quelled: the Bull, the Boar, I saw expire
> Under my hands: I seized the queenly Zone,
> And Diomede's fierce steeds I made my own.
> I plucked the golden Apples: Geryon slew:
> And what I could achieve Augeas knew:
> The Hind I caught: the vile Birds ceased their flight:
> Cerberus I dragged upwards; and gained Olympus' height.

The same 12 labors are listed, but again in a slightly different order, more than likely to fit the verse.

Apollodorus, Diodorus Siculus, Hyginus, and Euripides each tells the tale of Hercules and 12 labors, but their stories are not all are the same. The four main authors are discussed here in order of importance: Apollodorus is considered the most reliable source, while Euripides' version is most different from the remaining three, and although most authors acknowledge that Hercules fought with centaurs, only Euripides refers to the centaur battle as one of the Twelve Labors.

Apollodorus is often referenced as "Apollodorus of Athens," and it is believed that he was born around 180 BC. According to the *The Oxford Classical Dictionary*, Apollodorus lived in Alexandria, but he left there in approximately 146 BC, moving to Athens, where he spent the rest of his life. He had varied interests and was considered to be a great scholar. His *Library* was a study of ancient Greek heroic mythology. The oldest discovered copy of this book dates to the first or second century. His account of the Herculean myth is commonly accepted by most as the authority.

Diodorus Siculus, or Diodorus of Sicily, was born in Agyrium and is said to have flourished under Caesar and Augustus, at least until 21 BC. From approximately 60 to 30 BC, Diodorus wrote a history of the world entitled *The Library of History*, in 40 books.

Book 4 deals with Greek mythology, and the 12 labors of Hercules are discussed therein. According to the *The Oxford Classical Dictionary*, Diodorus' material depended greatly on the work of Apollodorus. Diodorus Siculus tried to combine the works of at least 15 other authors, with a result of confusion in the details. However, although the writings of Diodorus may seem undistinguished, repetitive, and contradictory, they are indeed of great value as they mention previous writers and authorities. Unfortunately, Diodorus borrowed freely from earlier authors, and modern scholars regard his works as unreliable.

There was a man named Hyginus who was a freedman of Augustus, appointed by Caesar to be the Librarian of the Palatine Library. This Hyginus was also a teacher and friend of Ovid. Regarding the mythological accounts of Hercules, a Latin work exists which is attributed to a Hyginus who cannot be identified with Augustus' freedman, according to the *The Oxford Classical Dictionary*. Hyginus wrote *Genealogiae*, a handbook compiled from Greek sources, outlining Greek mythology, and it was probably written in the second century. Unfortunately, it contains many errors, which, according to scholars, were caused by the writer's lack of knowledge of the Greek language.

Euripides is the earliest of the four main authors of Hercules. He was born in approximately 480 or 485 BC and died in early 406 BC. According to scholars, Euripides produced 22 tragedies and wrote 92 plays in total (80 titles are known). His version of the Labors of Hercules are outlined in a play entitled *The Madness of Hercules*, which takes place in Thebes, before the royal palace. An excerpt from his play, with a chorus entitled "The Lay of the Labours of Hercules," is provided in the next section. Euripides' account is not as factual as the others, although the verses do offer information from 300 years before the time of Apollodorus.

C.2. The Lay of the Labours of Hercules

Hard on the pæan triumphant-ringing
Oft Phoebus outpealeth a mourning-song,
O'er the strings of his harp of the voice sweet-singing

Sweeping the plectrum of gold along.
I also of him who hath passed to the places
Of underworld gloom—whether Zeus' Sons story,
Or Amphitryon's scion be theme of my
Sing: I am fain to uplift him before ye
Wreathed with the Twelve Toils' garland of glory:
For the dead have a a heritage, yea, have a crown,
Even the deathless memorial of deeds of renown.

I. The Nemean Lion

In Zeus' glen first, in the Lion's lair,
He fought, and the terror was no more there;
But the tawny beast's grim jaws were veiling
His golden head, and behind swept, trailing
Over his shoulders, its fell of hair.

II. The Centaurs

Then on the mountain-haunters raining
Far-flying arrows, his hand laid low
The tameless tribes of the Centaurs, straining
Against them of old that deadly bow.
Peneius is witness, the lovely-gliding,
And the fields unsown over plains wide-spreading,
And the hamlets in glens of Pelion hiding,
And on Homole's borders many a steading,
Whence poured they with ruining hoofs down-treading
Thessaly's harvests, for battle-brands
Tossing the mountain pines in their hands.

III. The Golden-horned Hind

And the Hind of the golden-antlered head,
And the dappled hide, which wont to spread
O'er the lands of the husbandmen stark desolation,
He slew it, and brought, for propitiation,
Unto Oenoë's Goddess, the Huntress dread.

IV. THE HORSES OF DIOMEDE

And on Diomede's chariot he rode, for he reined them,
By his bits overmastered, the stallions four
That had ravined at mangers of murder, and stained them
With revel of banquets of horror, when gore
From men's limbs dripped that their fierce teeth tore.

V. CYCNUS THE ROBBER

Over eddies of Hebrus silvery-coiling
He passed to the great work yet to be done,
In the tasks of the lord of Mycenae toiling;
By the surf mid the Maliac reefs ever boiling,
And by founts of Anaurus, he journeyed on,
Till the shaft from his string did the death-challenge sing
Unto Cycnus the guest-slayer, Amphanae's king,
Who gave welcome to none.

VI. THE GOLDEN APPLES

To the Song-maids he came, to the Garden enfolden
In glory of sunset, to pluck, where they grew
Mid the fruit-laden frondage the apples golden;
And the flame-hued dragon, the warder that drew
All round it his terrible spires, he slew.

VII. EXTIRPATION OF PIRATES

Through the rovers' gorges seaward-gazing
He sought; and thereafter in peace might roam
All mariners plying the oars swift-racing.

VIII. THE PILLARS OF HEAVEN

To the mansion of Atlas he came, and placing
His arms outstretched 'neath the sky's mid-dome,
By his might he upbore the firmament's floor,
And the palace with splendour of stars fretted o'er,
The Immortals' home.

IX. THE AMAZON'S GIRDLE

On the Amazon hosts upon war-steeds riding
By the shores of Maeotis, the river-meads green,
He fell; for the surges of Euxine he cleft.
What brother in arms was in Hellas left,
That came not to follow his banner's guiding,
When to win the Belt of the Warrior Queen,
The golden clasp of the mantle-vest,
He sailed far forth on a death-fraught quest?
And the wild maid's spoils for a glory abiding
Greece won: in Mycenae they yet shall be seen.

X. THE HYDRA

And the myriad heads he seared
Of the Hydra-fiend with flame,
Of the murderous hound Lernaean.

XI. THE THREE-BODIED GIANT GERYON

With its venom the arrows he smeared
That stung through the triple frame
Of the herdman-king Erythaean.

XII. CERBERUS

Many courses beside hath he run, ever earning
Triumph; but now to the dolorous land,
Unto Hades, hath sailed for his last toil-strife;
And there hath he quenched his light of life
Utterly—woe for the unreturning!
And of friends forlorn doth thy dwelling stand;
And waits for thy children Charon's oar
By the river that none may repass any more,
Whither godless wrong would speed them: and yearning
We strain our eyes for a vanished hand.
But if mine were the youth and the might

Of old—were mine old friends here,
   Might my spear but in battle be shaken,
I had championed thy children in fight:—
But mid desolate days and drear
I am left, of my youth forsaken!

According to scholars, in sections II, V, VII, and VIII of the above chorus, later writers substituted verses for the Erymanthian boar, the Augean stables, the Stymphalian birds, and the Cretan bull. However, I chose to share the original version here.

D

The Laplace Transform

D.1. Initial Value Problems and the Laplace Transform

D.1.1. Theory

Often we are faced with solving initial value problems that are differential equations with constant coefficients, but they may have a piecewise continuous or step forcing function. The forcing function is also known as a "driving" function since it comes from an external force or source that drives the system. One approach to solving such a problem is to transform the equation into another equation that is hopefully easier to work with. We then solve the new equation and reverse the transformation, interpreting the solution back in the context of the original problem. In particular, one method is to use Laplace transforms, which are used to transform the initial value problem into an algebraic equation that can then be easily solved using algebra.

The Laplace transform of a function $y(t)$ uses integration in comparing $y(t)$ to an exponential function of the form e^{st}. We define the Laplace transform function $\mathcal{L}[y](s)$ of the function $y(t)$ to be

$$\mathcal{L}[y](s) = \int_0^\infty y(t)e^{-st}\, dt. \qquad (D.1)$$

Mathematically, the Laplace transform is a function that converts $y(t)$ into a new function $Y(s)$, but we use the script letter $\mathcal{L}$ to represent that function (we usually write that $Y(s) = \mathcal{L}[y](s)$, or simply, $Y = \mathcal{L}[y]$. For example, if $y(t) = e^{3t}$, then its Laplace transform $\mathcal{L}[y](s)$ is determined by evaluating the improper integral

$$\mathcal{L}[y](s) = \int_0^\infty e^{3t} e^{-st}\, dt$$

$$= \int_0^\infty e^{(3-s)t}\, dt$$

161

$$= \lim_{b \to \infty} \int_0^b e^{(3-s)t} \, dt$$

$$= \lim_{b \to \infty} \left[\frac{-e^{(3-s)t}}{3-s} \right]_0^b$$

$$= \lim_{b \to \infty} \left[\frac{-(e^{(3-s)b} - 1)}{3-s} \right]. \tag{D.2}$$

The evaluation of the improper integral using the limit assumes that $s > 3$. If $s \le 3$, then the integral does not converge. Thus, we write

$$\mathcal{L}[e^{3t}] = \frac{1}{s-3} \qquad \text{for } s > 3.$$

Most calculus and differential equations textbooks have tables of commonly used Laplace transforms (of polynomials, of exponential and trigonometric functions, of step functions, and so forth). In addition, computer algebra systems, such as *Mathematica* and *Maple*, also have built-in algorithms for applying the Laplace transform operator. Here is a procedure for using Laplace transforms with initial value problems containing constant coefficients:

Step 1. Apply the Laplace transform to the initial value problem. This yields an algebraic equation in the transform of the solution. Any initial condition and forcing function is also transformed.

Step 2. Solve the algebraic equation for $\mathcal{L}[y](s)$, the transform. Use algebraic manipulation to ensure each term in the transform's formula is itself in the form of a Laplace transform (see the example).

Step 3. Apply the inverse Laplace transform to obtain the solution to the initial value problem.

The Laplace transform has a very special property involving its derivative that really forms the basis for its use in solving differential equations. Given a function $y(t)$ with associated Laplace transform

$\mathcal{L}[y]$, the Laplace transform of $\frac{dy}{dt}$ is

$$\mathcal{L}\left[\frac{dy}{dt}\right] = s\mathcal{L}[y] - y(0). \tag{D.3}$$

This is easily verified by using the definition of $\mathcal{L}\left[\frac{dy}{dt}\right]$ and evaluating the improper integral using integration by parts.

D.1.2. An Example

As an example, consider the initial value problem $\frac{dy}{dt} - 2y = \sin(3t)$, with $y(0) = 4$. Implementing Step 1, we apply the transform (writing $\mathcal{L}[y](s)$ as $\mathcal{L}[y]$):

$$\mathcal{L}\left[\frac{dy}{dt}\right] - 2\mathcal{L}[y] = \mathcal{L}[\sin(3t)],$$

$$s\mathcal{L}[y] - y_0 - 2\mathcal{L}[y] = \frac{3}{s^2 + 3^2},$$

where the Laplace transform of $\sin(3t)$ is found in a table. With Step 2, we solve for $\mathcal{L}[y]$:

$$\mathcal{L}[y] = \frac{y_0}{s-2} + \frac{3}{(s-2)(s^2+3^2)} = \frac{4}{s-2} + \frac{3}{(s-2)(s^2+3^2)}.$$

The partial fractions decomposition of the second fraction is

$$\frac{3}{(s-2)(s^2+3^2)} = \frac{3}{13}\left[\frac{1}{s-2} - \frac{2}{s^2+3^2} - \frac{s}{s^2+3^2}\right],$$

so

$$\mathcal{L}[y] = \frac{4}{s-2} + \frac{3}{13}\left(\frac{1}{s-2}\right) - \frac{2}{13}\left(\frac{3}{s^2+3^2}\right) - \frac{3s}{13}\left(\frac{s}{s^2+3^2}\right).$$

Using tables again, we obtain the inverse transform of each term on the right:

$$y(t) = 4e^{2t} + \frac{3}{13}e^{2t} - \frac{2}{13}\sin(3t) - \frac{3}{13}\cos(3t),$$

which is Step 3. Differentiating both sides easily verifies that this is indeed the solution.

Solution to the Sudoku Puzzles

E	R	Y	M	A	N	T	H	I
T	A	H	I	Y	R	M	N	E
I	N	M	E	H	T	Y	R	A
H	T	I	R	M	E	A	Y	N
Y	M	A	T	N	H	I	E	R
R	E	N	A	I	Y	H	T	M
A	H	E	N	T	I	R	M	Y
M	Y	R	H	E	A	N	I	T
N	I	T	Y	R	M	E	A	H

Figure E.1. Solution to the Erymanthian sudoku puzzle.

8	5	7	9	6	4	3	1	2
9	4	3	2	1	7	8	6	5
1	6	2	3	5	8	9	4	7
3	2	6	7	8	5	4	9	1
4	8	5	1	9	2	7	3	6
7	9	1	4	3	6	2	5	8
6	1	4	8	7	3	5	2	9
5	3	8	6	2	9	1	7	4
2	7	9	5	4	1	6	8	3

Figure E.2. Solution to the Diomedes sudoku puzzle.

1	2	7	3	5	4	7	6	7
3	7	5	6	1	7	2	7	4
4	7	6	7	7	2	3	1	5
5	1	7	7	2	3	6	4	7
7	6	3	4	7	1	7	5	2
7	4	2	5	6	7	7	3	1
2	3	4	1	7	7	5	7	6
6	7	1	2	3	5	4	7	7
7	5	7	7	4	6	1	2	3

Figure E.3. Solution to the Cerberus sudoku puzzle.

BIBLIOGRAPHY

Apollodorus. *Apollodorus, the Library*, with an English Translation by Sir J. G. Frazer, in 2 Volumes. Harvard University Press. Cambridge, Massachusetts. 1939. Includes Frazer's notes.

D. Arney, F. Giordano, and J. Robertson. *Mathematical Models with Discrete Dynamical Systems*, edited by A. Heidenberg and M. Huber. McGraw-Hill Primis Custom Publishing, New York. 2002.

W. W. R. Ball and H. S. M. Coxeter. *Mathematical Recreations and Essays*. 13th Ed. Dover Publications, Inc., New York. 1987.

D. Blatner. *The Joy of π*. Walker Publishing Company, Inc., New York. 1997.

C. H. Brase and C. P. Brase. *Understandable Statistics*. Houghton Mifflin Company, New York. 2006.

Cleere, J. Personal conversation, 13 February 2008. Dr. Cleere is Assistant Professor and Beef Cattle Specialist, Texas AgriLife Extension, Department of Animal Science, Texas A&M University, College Station, Texas.

T. Dantzig. *Mathematics in Ancient Greece*. Dover Publications, Inc., Mineola, New York. 2006.

Diodorus of Sicily. *Diodorus Siculus*, with an English Translation by C. H. Oldfather, in 10 Volumes. Harvard University Press. Cambridge, Massachusetts. 1935. Includes Oldfather's notes.

The Euler Archives, found online at http://www.math.dartmouth.edu/~euler/pages/E476.html. Accessed July 2007.

167

Euripedes. *Euripedes*, with an English Translation by A. S. Way, in 4 Volumes. G. P. Putnam's Sons Publishing, New York. 1925. Includes Way's notes.

T. Farmer and F. Gass. "Physical Demonstrations in the Calculus Classroom." *The College Mathematics Journal.* Vol. 23, No. 2 (March 1992), pp. 146–148.

S. Garfunkel, project director. *For All Practical Purposes: Mathematical Literacy in Today's World.* 7th Ed. W. H. Freeman and Company, New York. 2006.

R. Graham, D. Knuth, and O. Patashnik. *Concrete Mathematics.* Addison-Wesley Publishing Company, Reading, Massachusetts. 1990.

The Greek Anthology, with an English Translation by W. R. Paton, in 5 Volumes. G. P. Putnam's Sons Publishing, New York. 1926. Includes Paton's notes.

C. W. Groetsch. "Inverse Problems and Torricelli's Law." *The College Mathematics Journal.* Vol. 24, No. 3 (May 1993), pp. 210–217.

T. L. Heath. *The Works of Archimedes.* Dover Publications, Inc., Mineola, New York. 2002.

D. Hughes-Hallett et al. *Calculus.* 4th Ed. John Wiley & Sons, Inc., New York. 2005.

Josephus. *The Jewish War*, with an English Translation by H. St. J. Thackeray. G. P. Putnam's Sons Publishing, New York. 1927. Includes Thackeray's ·notes.

E. R. Kandel, J. H. Schwartz, and T. M. Jessell. *Principles of Neural Science.* 4th Ed. McGraw-Hill, New York. 2000.

M. Kraitchik. *Mathematical Recreations.* 2nd Rev. Ed. Dover Publications, Inc., New York. 1953.

The Let's Make a Deal Applet. Found online at http://www.stat.sc.edu/~west/javahtml/LetsMakeaDeal/html. Accessed May 2007.

The Monte Carlo JAVA Applet. Found online at http://polymer.bu.edu/java/java/montepi/montepiapplet.html. Accessed June 2007.

J. P. Morgan, N. R. Chaganty, R. C. Dahiya, and M. J. Doviak. "Let's Make A Deal: The Player's Dilemma." *The American Statistician.* Vol. 45, No. 4 (Nov. 1991), pp. 284–287.

The Oxford Classical Dictionary. S. Hornblower and A. Spawforth, editors. Oxford University Press, New York, 1996.

Perseus Digital Library Project. Ed. G. R. Crane. March 16, 2000. Tufts University. Accessed January 2007. *http://www.perseus.tufts.edu.*

B. B. Powell. *Classical Myth*. 5th Ed. Pearson Education, Inc., Englewood Cliffs, New Jersey. 2007.

Professional Bull Riders, Inc. Home Page. Found online at http://www. pbrnow.com. Accessed July 2007.

J. K. Rowling. *Harry Potter and the Sorceror's Stone*. Scholastic, Inc., New York. 1997.

"They Said It." *Sports Illustrated*. Vol. 91, Issue 3 (July 19, 1999), p. 35.

INDEX